Doing Security

Crime Prevention and Security Management

Series Editor: **Martin Gill**

Titles include:

Mark Button
DOING SECURITY
Critical Reflections and an Agenda for Change

Kate Moss
SECURITY AND LIBERTY
Restriction by Stealth

Forthcoming

Joshua Bamfield
SHOPPING AND CRIME

Paul Ekblom
THE 5IS FRAMEWORK FOR CRIME PREVENTION AND COMMUNITY
SAFETY

Bob Hoogenboom
GOVERNANCE AND POLICING OF SECURITY
Exploring the Shifting Contours of Control

Crime Prevention and Security Management
Series Standing Order ISBN   978–0–230–01355–1 hardback
                                978–0–230–01356–8 paperback
                                (*outside North America only*)

You can receive future titles in this series as they are published by placing a standing order.
Please contact your bookseller or, in case of difficulty, write to us at the address below with
your name and address, the title of the series and the ISBN quoted above.

Customer Services Department, Macmillan Distribution Ltd, Houndmills, Basingstoke,
Hampshire RG21 6XS, England.

Doing Security

Critical Reflections and an Agenda for Change

Mark Button
University of Portsmouth, UK

First published 2008 by
PALGRAVE MACMILLAN

Palgrave Macmillan in the UK is an imprint of Macmillan Publishers Limited,
registered in England, company number 785998, of Houndmills, Basingstoke,
Hampshire RG21 6XS.

Palgrave Macmillan in the US is a division of St Martin's Press LLC,
175 Fifth Avenue, New York, NY 10010.

Palgrave Macmillan is the global academic imprint of the above companies
and has companies and representatives throughout the world.

Palgrave® and Macmillan® are registered trademarks in the United States,
the United Kingdom, Europe and other countries.

ISBN-13: 978–0–230–55311–8   hardback
ISBN-10: 0–230–55311–7   hardback

This book is printed on paper suitable for recycling and made from fully
managed and sustained forest sources. Logging, pulping and manufacturing
processes are expected to conform to the environmental regulations of the
country of origin.

A catalogue record for this book is available from the British Library.

A catalogue record for this book is available from the Library of Congress.

10   9   8   7   6   5   4   3   2   1
17 16 15 14 13 12 11 10 09 08

Printed and bound in Great Britain by
CPI Antony Rowe, Chippenham and Eastbourne

*This book is dedicated to my mother, Sally
and my late father, Andrew*

Contents

List of Figures

List of Tables and Boxes

Tables

Boxes

Series Editor's Preface

In a series that has evolved to generate new insights, this book makes an important contribution to the much neglected area of private security generally and contemporary security management in particular. That it is written by Mark Button who has established himself as one of the world's leading scholars in this field of study adds to the interest. In this text he critiques the contribution that the private security sector has made to crime prevention and finds it wanting. He highlights examples of security failing to meet reasonable requirements in a variety of settings, sometimes leading to disaster. And his summary of how offenders make decisions leads him to conclude, surely rightly, that all too often measures are easily defeated by even the most rudimentarily prepared adversary. In assessing both the technical and human elements of security Mark highlights its shortcomings and provides a basis for presenting a framework for an 'agenda for change'.

His analysis leads him to adopt a wide-ranging critique. This includes a focus on the issues of the regulation of security, not least the gaps in current provision; the inadequacies of the security infrastructure; and security inequity. He refers to these as the 'foundations of security' and in each case he presents an alternative model for more effective and equitable security management.

Never far from his assessment is the role of humans in providing effective security. Fuelled by a range of case studies which show that security measures fail because of the human element, he presents his own views on the problems and potential solutions. He argues for a redefinition of 'security management' as 'security risk management', and his roadmap for achieving professional status – the key to improving the human element in security – is laudable and ambitious.

Mark has not shied away from tackling some of the main concerns – certainly from some academic commentators – about the role of security in society. For example, the Achilles heel of private security, certainly when discussed in the context of providing a public good, is that it is accountable to those who pay for it, and this, logically, is rarely the poor. And so he promotes the idea of security unions based on cooperative principles, to redress inequities.

In this book Mark covers a lot of ground and does so at a number of levels. On the one hand he draws upon (and sometimes adapts) knowledge from a range of conceptual frameworks, including situational prevention, and the work of Lukes and his three dimensions of power. On the other he deals with practicalities such as the appropriate content for the training of security officers and the education of security managers. And he draws upon evidence from across the world.

Mark has advocated a new way forward that will inevitably be seen as controversial by some and as such there is much to argue about and potentially disagree with. That perhaps is the sign of a worthwhile text. What is in little doubt is that private security generally, and security management in organisations included, have much untapped potential to contribute further to the prevention of crime. The effectiveness of what emerges in future years will depend in part on the quality of informed debate now, and in setting out an 'agenda for change' Mark has made an enormously important contribution.

MARTIN GILL

Acknowledgements

As with any project of this nature there are too many people to thank and someone is always forgotten. Nevertheless there are some who deserve particular mention. First I would like to thank Azeem Aleem, Alison Wakefield, Mike Tiller and Les Johnston for all offering extensive comments on drafts of this book. Some of the ideas for the book were debated in University of Portsmouth seminars and the contributions of Steve Savage, Francis Pakes, Carol Hayden and Mike Nash must also be acknowledged. Bruce George as ever has provided encouragement and stimulated ideas. I am also grateful for Jorma Hakala for inviting me to Finland to show me a model security industry and regulatory body. Many of the reforms I propose in this book have been implemented there for some time. I would also like to thank Martin Gill for his comments on the original proposal and in particular his insistence that I needed 80,000 words for the project, he was very right, and I was wrong. He must also be thanked for his comments on the first draft. The staff of Palgrave have also been central to achieving this project and they must also be thanked. Finally I would like to thank my partner Emma and my son Conrad for persevering with my endless discussions and late night writing sessions completing this book.

List of Abbreviations

ACPO	Association of Chief Police Officers
ASIS	American Society for Industrial Security
BRC	British Retail Consortium
BS	British Standard
BSIA	British Security Industry Association
CAPPS	Computer assisted passenger pre-screening system
CCTV	Close circuit television
CDRP	Crime and Disorder Reduction Partnerships
CIPD	Chartered Institute of Personnel and Development
CoESS	Confederation of European Security Services
CPTEDM	Crime Prevention Through Environmental Design Management
CV	Curriculum vitae
CVIT	Cash and valuables in transit
EAS	Electronic article surveillance equipment
FAA	Federal Aviation Administration
FMECA	Failure modes and effects criticality analysis
FSA	Financial Services Authority
GSO	General Security Officer
GSOSv	General Security Officer Supervisor
HRM	Human resource management
IIS	International Institute of Security
ISM	Institute of Security Management
ISMA	International Security Management Association
NHS	National Health Service
NHSCFSMS	National Health Service Counter Fraud and Security Management Service
NQF	National Qualifications Framework
NSI	National Security Inspectorate
PIRA	Provisional Irish Republican Army
POA	Prison Officers Association
PSIA	Private Security Industry Act
PSO	Public Security Officer
PSOSv	Public Security Officer Supervisor
QRA	Quantified risk analysis
ROI	Return on investment

RUC	Royal Ulster Constabulary
SBD	Secured by design
SIA	Security Industry Authority
SITO	Security Industry Training Organisation
SPAD	Signal passed at danger
SRM	Security risk management
SyI	The Security Institute
TRANSEC	Transport Security Division
TSO	Trainee Security Officer

Introduction

Security is an important issue. It is important to a domestic householder who doesn't want to be burgled, to a corporation suffering losses from staff pilferage, to governments fearing terrorist attacks. The 'security systems' that are created to counter these and many other risks vary significantly in their complexity, in their effectiveness and in their implications (Johnston and Shearing, 2003; Gill, 2006). There are, nevertheless, some recurring themes and paradoxes concerning security (Zedner, 2003a; Loader and Walker, 2007). Security is generally held to be of paramount importance, yet the quality of security provision in many organisations and areas is often poor (House of Commons Defence Committee, 1990; House of Commons Home Affairs Committee, 1995; Gill, 2006; Button, 2007a). Security systems frequently fail. Elements of security systems are routinely derided, most commonly security officers (Shearing et al., 1985; Button, 2007a). And governments often take a limited interest in enhancing security in the private sector and even in much of the public sector (Button and George, 2001). The delivery of security is frequently delegated to personnel with limited training, inadequate education and no real commitment to professionalism (Parfomak, 2004). To cut straight to the point, there are some fundamental problems hindering the effective development of security in society as a whole and at an organisational or community level. Too much security, to use an increasingly popular phrase, is simply 'sub-prime'. The time has therefore come to provide a frank critique of what is wrong with the delivery of security and then to set out an agenda for change. *Doing Security* seeks to undertake this challenge, and in line with other volumes in Palgrave Macmillan's 'Crime Prevention and Security Management Series', it also aims to stimulate new thinking, provide new insights and adopt new perspectives on security.

This book will set out an agenda for the transformation of security by creating a model for developing the most effective security system for various contexts, from a housing estate and shopping centre right up to the national level. It also seeks to provide a framework for a security system that actually minimises the need for security interventions and when they are required, maximises their effectiveness. In particular, it seeks to showcase the huge benefits of an understanding of research findings from disciplines as diverse as psychology, criminology, sociology and geography for the security practitioner.

This book will also advocate some national-level reforms required to enhance the effectiveness of security by rebuilding its foundations. Some of the changes advocated are drawn from a burgeoning academic literature on security, which sets out some exciting theoretical frameworks and ideas for structural reform with the aim of delivering effective security. Shearing and others have set out a framework for assessing the governance of security using a 'nodal model' (to be defined in Chapter 1) as well as radical ideas for combating security inequity (Johnston and Shearing, 2003; Wood and Shearing, 2007). Their proposals and ideas do not only provide material for academic debate, but also comprise a body of work, largely ignored by policymakers and security practitioners, for the practical enhancement of security. This book will therefore attempt the precarious task of bringing some of this rich vein of theoretically-inspired frameworks and ideas to security practitioners. It will seek to provide a bridge from academia to the largely scholar-less security community.

Doing Security is aimed at security managers, those with responsibility for security (such as the military, police, customs officers and so on), students and researchers in the security field, and policymakers with a security remit. *Doing Security* is not a text book on how a security manager should do security in a particular context. There are lots of those types of books on the market (Fischer and Green, 1988 for example). It is more than that. It will provide a strong critique of those who manage security. It will show that they are both the barrier to change and the most significant part in the chain to enhancing security. It will argue that their discipline has reached a Rubicon and that it can either embrace an agenda for change and go through the pain of becoming the profession that so many want, or it can continue to be lumped together with other quasi-professional service functions, such as cleaning and facilities management – thus failing to provide the contribution to the enhancement of security that is needed.

The book uses the experience of the United Kingdom as the basis for much of its critique and agendas for change. However, it is not intended for this book to provide a blueprint of action for the UK alone. Many of the problems outlined in this book will be familiar to readers from across the globe. Indeed in identifying a way forward for 'Doing Security' the book also draws upon best practice from around the world. There are many examples of innovation and best practice which – if only they could be stitched together in one nation – could produce a remarkable improvement in security.

Organisation of the book

Doing Security is divided into three main parts, with conclusions presented in a fourth. Part I is entitled 'Mapping Security' and contains only one chapter, which explores the expanding size and role of the private security industry in the delivery of security. It also examines what are emerging as the four main perspectives on security and in particular the growing school of thought centred upon nodal governance. The chapter seeks to add further flesh to the bones of these ideas by filling some existing gaps and expanding upon underdeveloped areas.

Part II is entitled 'Security Undone' and contains four chapters undertaking a critique of the effectiveness of security. Chapter 2 explores the neglected subject of security failure. Evidence is used to demonstrate the large extent of security failure in society and the very different consequences of such failures. The much more widely researched fields of accidents and disasters are examined to show the relevance of some of the theories for explaining security failures. The Gardner Museum heist, the 11 September attacks and Prince William's twenty-first birthday are utilised as case studies to illustrate this. From this analysis common causes of security failure are modelled.

Chapter 3 looks at malefactors and their decision-making strategies, with a particular focus on their views of security. It highlights the low regard that many malefactors have for security and that it is more often than not considered no more than a hurdle to overcome. It also profiles the diverse range of different types of malefactors who may target nodes, from opportunistic workers and organised crime networks to terrorists. The subject of the third chapter in Part II is the human element in the security system: security officers and security managers. It will show how there are inherent weaknesses in their occupational cultures that undermine their effectiveness.

The final chapter in this part moves on to consider the foundations of security. It shows that there are structures in place that naturalise the underperformance of the human element. From illustrating some of the inadequacies in the regulatory structure for private security as a whole, the chapter moves on to highlight the weaknesses in the professional infrastructure of security practitioners. It also explores the issue of inequity in security provision and how that too contributes to increased insecurity.

Part III of the book moves on to 'Doing Security' and sets out an agenda for reforming security. The first of the four chapters in this part of the book offers a description of how a holistic model for developing a security system can be developed. Chapters 7 and 8 draw upon Lukes's (1974) theoretical framework for deconstructing power into three dimensions. Chapter 7 identifies third-dimension security strategies that can be used to create structures that make malefactors less likely to target a specific node or to engage in deviant behaviour. The following chapter focuses upon second- and first-dimension security strategies – in particular the human element – and how mere presence can achieve outcomes and enhance security. It also provides insights for enhancing security when intervention is required by security officers.

The final chapter of Part III examines how the foundations of security can be rebuilt. It shows how regulation can be enhanced to improve security, provides a route map to enhancing the professional infrastructure and in so doing argues for the reconfiguration of security management as Security Risk Management (SRM). The chapter ends with radical proposals for security unions as a means to address security inequity.

In Part IV, the concluding chapter brings together the main ideas for 'Doing Security' to create one model. The chapter also identifies some areas requiring further research and explores the neglected issue of the role of social decay in fuelling insecurity.

Part I
Mapping Security

1
Private Security, Nodal Governance and the Security System

Introduction

Security – or insecurity – has become a major public concern, an issue at the top of the political agenda and the subject of expanding academic enquiry. Fuelled by the rise of new terrorism, fears exacerbated by high profile crimes and growing anxieties over anti-social behaviour, for many people the need for more effective security has never seemed greater. In recent years there has been a proliferation of books and articles dedicated to security, examining governance, developing theory and identifying best practice in a variety of areas (Johnston and Shearing, 2003; Loader and Walker, 2007; Gill, 2006). There are now dozens of academics, worldwide, researching and writing on the subject of security. One of the most significant areas of interest to emerge has been around the debates relating to nodal governance. This is the main topic of this chapter and its exploration here will provide a theoretical framework for the rest of the book. Before some of these discussions are considered, however, it is important to grapple with some important foundational concepts and issues, such as what is meant by security, the growing plurality of bodies involved in delivering it, the remarkable growth of private security and the reactions to that expansion.

Security, pluralisation and the rise of private security

To begin with it is worth defining what is actually meant by the term 'security'. Neither security – nor its opposite, insecurity – are objective or measurable feelings; they are, as Wood and Shearing (2007) argue, 'imagined'. 'Security' is used to cover a much wider range of conditions than the most salient examples relating to crime. It is used in international relations, and in relation to welfare payments. In French, *sécurité* is used even more broadly, indicating safety as well as security in the English sense (Gill, 1996). Security is also a 'natural' process. There are many examples in the natural world of animals, plants and even viruses developing security tactics by pursuing certain behaviours or even by building security structures (Ekblom, 1999). Possibly the best capture of the essence of security is by Zedner (2003b: 155, cited in Wood and Shearing, 2007: 4).

> Security is both a *state of being* and a means to that end. As a state of being, security suggests two quite distinct objective and subjective conditions. And as an *objective condition*, it takes a number of possible forms. First, it is the condition of being without threat: the hypothetical state of absolute security. Secondly, it is defined by the neutralization of threats: the state of 'being protected from'. Thirdly, it is a form of avoidance or non-exposure to danger... As a *subjective condition*, security again suggests both the positive condition of feeling safe, and freedom from anxiety or apprehension defined negatively by reference to insecurity.

It is also important to recognise the growing body of work that sees security in a much broader framework, as 'human security' (Commission on Human Security, 2003). The UN Commission examined a wide range of insecurities based in legal, environmental, economic, educational and health contexts as well as in 'traditional' security areas and explored their links to one another. As Kofi Annan argued,

> Human security in its broadest sense embraces far more than the absence of violent conflict. It encompasses human rights, good governance, access to education and healthcare and ensuring each individual has opportunities and choices to fulfil his or her own potential. Every step in this direction is also a step towards reducing poverty, achieving economic growth and preventing conflict. Freedom from want, freedom from fear and the freedom of future generations to inherit a healthy natural environment – these are the interrelated building blocks of human, and therefore national, security.
>
> (Commission on Human Security, 2003: 4)

Some aspects of these ideas on enhancing human security will be developed further in chapters 5 and 9.

Traditional conceptions of the delivery of security have centred around the state as the primary delivery mechanism. However, a growing body of evidence demonstrates that the governance of security has become much more nodal, with 'pluralised' or 'fragmented' modes of delivery (Bayley and Shearing, 1996; Johnston, 2000; Button, 2002; Van Stedan, 2007; White, forthcoming). A number of writers have illustrated the ways in which state police are supported by hybrid bodies such as trading standards officers, environmental health officers and benefits fraud investigators; charitable organisations such as the Royal Society for the Prevention of Cruelty to Animals (RSPCA); voluntary bodies such as the Special Constabulary; and vigilante groups such as the Guardian Angels (Johnston, 1992; Button, 2002; Crawford et al., 2005). Central to this shift has also been the rise in function and size of the private security industry across many jurisdictions (Cunningham et al., 1990; Jones and Newburn, 1998; Button, 2007a).

Internationally there is no security function carried out in the state sector that is not undertaken in the private sector in some form. For example, Jones and Newburn (1998) in an analysis of selected policing functions, such as responding to calls, investigating crimes, arrest of offenders and so on, found that the Metropolitan Police and the private security industry carried out the same wide range of selected tasks. Virtually every function carried out by the police was offered as a service by the private sector in a similar form. In the UK, probably the only functions not carried out by the private sector relate to firearms, but in many other countries private security staff are routinely armed (Cunningham et al., 1990; De Waard,1993; Rigakos, 2002; Sarre and Prenzler, 2005). In some countries where there are serious levels of gun violence private security firms offer armed response services. For example, in South Africa ADT Security has a fleet of over 600 vehicles in 386 suburbs ready to respond to an incident 24 hours a day (ADT Security, n.d.) Perhaps the greatest illustration of the capacity of the private sector to undertake almost any role has been the growing involvement of private contractors in Iraq. Here between 6,000 and 20,000 overseas security personnel are estimated to be engaged in the protection of foreign workers and facilities, providing advice as well as logistics and technical support. Much of their work involves functions traditionally associated with the military and the high financial rewards of undertaking such work are accompanied by high casualty rates (Johnston, 2006).

A lot of statistical research has been undertaken in order to esti-mate the numbers of security officers employed in a country, usually in relation to the police. However, there are a variety of problems to be overcome in seeking to gauge the number of security officers, for example, defining the police and security, methods of data collection and the large numbers of part-time security officers (see Stenning, 1994; Van Steden and Sarre, 2005; Button, 2007b). Bearing these caveats in mind, several studies have shown that the extent to which the numbers of security officers have grown is such that in some countries they now outnumber police officers. Recent research on the European Union (EU) (Button, 2007b) identified Hungary as having the highest security officer to police officer ratio at 2:1. In Poland it was 1.94:1, in Ireland 1.67:1 and the United Kingdom (UK) it was 1.06:1. Most EU countries employed more police officers than security officers with an EU average of 0.73 security officers to police officers. In North America research generally suggests that there are more security officers than police officers, with Canada having two security officers to every police officer (Rigakos and Yeung, 2006). Similarly in the USA a figure of 2.6 security staff to every police officer has been suggested by Cunningham et al. (1990), a ratio that – given the date of publication – might well be even greater now. In Australia, Prenzler and Sarre (2006) cite census data to suggest that by 2010 the number of security officers will exceed police officers. The emerging theme, bearing the caveats in mind, is of growing numbers of security officers that in several countries are comparable to or exceed the number of police officers.

The range of roles undertaken and the increasing size of the private security industry have fuelled growing debates over the orientation of this sector. Out of this, four perspectives can be distinguished on the rise of private security; three of them have antecedents in the three main perspectives explaining the growth of private security noted by Jones and Newburn (1998).

Radical negative

The radical negative perspective is antipathetic to the emergence of private security and would restrict or prevent its growth either through restrictive regulation or other measures. Much of the literature associated with this is linked to American industrial disputes in the late nineteenth and early twentieth centuries. The radical negative perspective views the growth of private security as an inevitable consequence of the crisis of capitalism wherein the state utilizes the private sector to strengthen its legitimacy (Spitzer and Scull, 1977; Weiss, 1978; Couch, 1987). In nineteenth and early twentieth-century America, the coal, iron and steel industries were developing in rural areas where the state was not well established. Industrial militancy threatened the corporations of the time and companies resorted to private security to maintain control. These private security services went well beyond keeping the peace and sought to ensure the obedience of the working classes. The companies drew upon their own private forces, but also made use of contractors for investigations and policing industrial disputes – most famously Pinkerton. Central to these negative views is an understanding of private security as representing a 'hard' approach to security, it emphasises their threat to the state, political motivation, confrontational style and reactive approach (Button, 2004).

Companies like Pinkerton provided a variety of services from general property protection services to a range of strike-breaking services such as labour espionage, strike-breakers, strike-guards and strike missionaries (those who were paid to persuade strikers to go back to work) (Weiss, 1978). The policing practices that emerged, particularly during strikes in this period, led to serious confrontations between the forces of capital and labour. There were many disputes where the civil rights of striking workers were violated and where some were even killed. Such were the problems throughout this period that 'private police systems' were the subject of many Congressional reports (see United States Committee on Education and Labour, 1971 – a reissue of a 1931 report). As the 1931 report argued:

> The use of private police systems to infringe upon the civil liberties of workers has a long and often blood stained history. The methods used by private armed guards have been violent. The purposes have usually been to prevent the exercise of civil rights in the self-organisation of employees into unions or to break strikes either called to enforce collective bargaining or to obtain better working conditions for union members.

A number of writers on policing in the UK during the 1970s were also critical of private security. Bowden (1978: 254) called private security forces 'private armies' and cited the former Labour Party leader George Lansbury who regarded the industry as the 'the first halting step towards fascism'. Bowden was concerned with the political activities of the industry, particularly those of private detectives investigating trade unionists and political activists. He wrote,

> Of particular concern is the fact that private police forces have been used, and will be used in the future, in Britain, for covert political

operations. Their powers of investigation and surveillance are considerable and have been given a boost in terms of information resources that they have available by the recruitment of senior ex-policemen, politicians and even Intelligence specialists onto their staffs.

(Bowden, 1978: 256)

Bunyan (1976) took a similar view and highlighted investigators involved in the gathering of information on political extremists and trade unionists for the 'Economic League' (an organisation dedicated to securing information on such groups for the business community). One former private detective from this period, Gary Murray, has illustrated in his memoirs how investigators were used not just by private companies to secure information on political activists and trade unionists, but also by foreign governments to secure intelligence on their political opponents in the UK, most prominently by the South African government on anti-apartheid protesters. He also offers evidence that the Security Service (MI5) subcontracted some of its work to private investigators (Murray, 1993).

More recent critics of the private security industry have argued that private security is 'paradoxical' and an 'oxymoron' (Zedner, 2003a; Loader, 1996). Loader argues that the provision of private security may actually exacerbate concern about crime with the consequence that it might 'institutionalise anxiety' (1996: 156). He also argues that the retreat behind fortresses of security serves to reinforce perceptions of danger beyond the protected area. Finally, he argues that private security may also be criminogenic by displacing crime to less protected areas and 'hollowing out' the less populated public space with crime and fear of crime becoming associated with such places. Zedner offers a similar perspective on private security based upon six paradoxes. Central to their analyses is the emergence of private security as a commodity, rather than as what was regarded traditionally as a public good delivered by the state police (Loader, 1999).

However, much of their critique is focused upon a very narrow part of the private security industry: firms delivering patrolling services on public streets and private security in gated communities. In the UK at least, these constitute a small part of the security market. The state has never provided burglar alarms to homeowners, provided security to factories or protected art galleries. The bulk of the security market is focused upon corporate and public sector clients, not members of the public.

Possibly the most radical attack on the private security industry is that of Sklansky (2006) who argues that the expansion of private security poses a threat to democracy. He argues, first, that the growth of private security will lead to inequity in security provision. Second, he sets out the implications of privatisation for the police themselves, arguing that attempts – however ineffective – at improving police legitimacy and accountability will be lost in their attempts to compete with the private sector and embrace increasingly 'market-oriented' mentalities. Finally he considers private security firms themselves to be anti-union, focused upon efficiency with little regard for rights and democracy.

Each of these attacks can be countered. The potential for inequity can be addressed through a redistribution of security, which will be explored later in this chapter. Competition between the public and private sectors does not have to mean the diminution of police commitment to democratic governance and rights or the pursuit of performance objectives taking precedence over public demands. What is needed is the establishment of appropriate regulatory frameworks. This can be done successfully as the work of Harding (1997) shows on prison privatisation, where the private sector can be regulated to provide more effective governance than the public sector. Finally, although there are undoubtedly 'bad' security companies with poor employee relations, focused upon narrow objectives with little commitment to rights and governance, there are also many companies that are the opposite to this. Further appropriate regulation can be pursued to limit these negative traits.

The earlier radical views owe much to the specifically American social changes and political controversies of the time. There are examples of private security involved in political disputes and defending private interests elsewhere, but not on the scale of the American examples or with such violence and controversy (Button and John, 2002). Similarly fears over 'spying' on political opponents by private investigators, at least in the UK, has waned with the decline of the hard left, though it does still take place, particularly vis-à-vis extreme animal rights activists and environmentalists.

Conservative negative

The conservative negative approach is based on the fear that private security might threaten certain established interests. It is most commonly associated with the police in the early stages of their relationship with the private security industry, which Stenning (1989) describes as based initially upon the 'denial of existence' of private security, followed by 'grudging recognition' and 'denigration' and finally 'competition' and 'open hostility'. There is much evidence throughout the world of the police and other state bodies opposing the expansion of private security, largely driven by the concern to defend their own interests.

In Australia in the build up to the Asia Pacific summit in Sydney in 2007 the plans to draft in private security to support the police were dropped following opposition by the police over concerns relating to the character of many private security staff (*Sydney Morning Herald*, 16 July 2007). Police unions and their equivalents have generally opposed the expansion of the role of private security into areas that threaten their activities, for example, the England and Wales Police Federation's opposition to private security street patrols and the Prison Officers' Association (POA) opposition to prison privatisation (Broughton, 1995; Prison Officers Association, 1997). Indeed the POA is a key member of the 'Public Services Not for Profit Campaign' which is a coalition, largely of trade unions, generally opposed to the privatisation of public services (POA, 2006).

These attitudes can be interpreted as the police encouraging the state to pursue restrictive regulations on the commercial sector; in some countries, however, state police forces at the same time actively compete with the private sector. For example in Georgia in the former Soviet Union the state police have their own guard force which is armed and competes for contracts. The commercial sector is not allowed firearms and in a country where firearms are a necessity this puts them at a strong disadvantage in the market. The police rationale for this is the need for revenue-raising mechanisms to supplement resources.

Another side to this perspective is the fear of how private security might evolve. For instance, the early debates over regulation of the private security industry in the UK during the 1960s and 1970s were won by the opponents of regulation driven by a perspective that regulation might give private providers a respectability and an authority they did not deserve. Proposals for the regulation of private investigators were rejected on the grounds that it might give them a 'licence to pry' (*Hansard*, 13 July 1973: col 1966). Similarly in 1981 Lord Willis's Security Officers Control Bill was opposed by some in the House of Lords on the grounds that it might give the public the impression that security officers have 'special powers' like those of the police (*House of Lords Report*, 3 December 1981: col 1166). Partly linked to these arguments has been a fear amongst some on the left of the Labour Party that the greater legitimacy that regulation might bring might enable further privatisation of the criminal justice system.

Although there are legitimate concerns over the expansion of private security this perspective offers nothing but a narrow defence of sectional interests. The state is not always the best at delivering services and neither is the private sector. There is much to learn from both sides. In the increasingly globalised environment and in a context of growing pressures on state resources it is only natural to look at more efficient and effective mechanisms of delivery. It is only to be expected that representative associations will seek to defend their interests, but what they are defending may not be in the broader public interest.

Traditional positive

The traditional positive perspective has its roots in the fiscal constraint, 'liberal democratic' perspective identified by Jones and Newburn (1998). The 'liberal democratic' view essentially interprets the growth of private security as a consequence of increasing demands on public police, which cannot be satisfied. As a result the private sector has stepped in to fill the gap. This is seen as a positive development, with varying views on the degree of state intervention required to shape the emergence of the industry in order for it to reach its full potential.

The first and most significant study associated with this perspective was Kakalik and Wildhorn's (1972 a, b, c, d, e) research, initially published by the Rand Corporation in 1971. This illustrated the significant role of the private security industry in preventing crime against corporations. It also emphasised the supportive and complementary role played by the private sector in relation to the public police and developed the idea of the industry as a 'junior partner' to the police in 'policing'. Until the publication of this study the private security industry had either been viewed as a potential 'private army' or dismissed as unimportant (Shearing, 1992). Kakalik and Wildhorn's research turned these ideas on their head, promoting a view of private security as a positive asset to society that could be more fully developed and utilised. This study was accompanied during the early 1970s by a range of further research published in the UK and USA which illustrated the growing importance of private policing (Braun and Lee, 1971; Scott and MacPherson, 1971; Wiles and McClintock, 1972; National Advisory Council on Criminal Justice Standards and Goals, 1976; Draper, 1978).

The US government funded yet another major study into the private security industry, this time undertaken by the National Advisory Council on Criminal Justice Standards and Goals and published in 1976. This lengthy report took up the ideas of the Rand study and the 'junior partner' view, seeking to establish common standards across the private security industry, which would improve its effectiveness and efficiency and therefore its contribution to policing. In the opening preface to the report the authors write:

> The application of the resources, technology, skills, and knowledge of the private security industry presents the best hope available for protecting the citizen who has witnessed the defenses [sic] against crime shrink to a level which leaves him virtually unprotected.

In the mid-1980s another US government-funded project was published that took the Rand ideas even further by advocating 'equal partnership'. The Hallcrest reports undertaken by Cunningham and Taylor (1985) and Cunningham et al. (1990) argued that the private security industry was a 'massively under-utilized resource' and went on to suggest that it had become the nation's 'primary protective resource'. They set out an analysis that advocated an increased role for the private sector in crime fighting as well as in crime prevention and for greater cooperation between public and private sectors. Further studies over the last decade have continued to emphasise the liberal democratic perspective. In the USA, research stimulated by Hallcrest into police moonlighting as private security operatives found that chiefs of police defended the practice on the grounds that police thus employed were simply doing what they were supposed to be doing, but at private expense (Reiss, 1988).

The dominant perspective on private security in the UK and North America would appear to be 'traditional positive'. Governments are seeking to expand the policing and crime prevention infrastructure in the face of accelerating crime rates in order to provide greater security in society. Private security has come to be seen as a positive resource to be developed and improved, with active partnerships encouraged between the public and private sectors. Some in the public sector see it as an additional resource to draw upon in order to achieve their goals and as a potential future career. Of concern in this perspective, however, is the lack of attention given to the broader implications of the growth of private security. Issues such as accountability, civil liberties and, most significantly, inequity in provision of security are neglected (Shearing, 1993; Johnston and Shearing, 2003). The culmination of this critique has been the emergence of the nodal perspective.

Nodal positive

The nodal analysis of security will be outlined in much greater depth below. In essence, however, it is based on an image of the dispersal of governance to a network of nodes from the state, corporate and voluntary sectors. In this perspective the benefits of the expansion of private security are recognised, but alongside this there is also concern over some aspects of private security operations, most significantly inequities in the provision of security. Central to this perspective is the advancement of a theoretical model of security delivery that seeks to redistribute security to nodes with security deficits. We will now consider this analysis in greater depth.

Nodal governance

The growing school of thought centred around the concept of 'nodal governance' challenges state-centred conceptions of the governance of security, recognising instead that the state constitutes one of the many nodes of a network in delivering security (Johnston and Shearing, 2003; Shearing and Wood, 2003). Security governance, it is argued, is also delivered in the corporate sector, non-government organisations and the voluntary and informal sectors. There are two parts to the nodal governance discourse. First is a theoretical analytical framework which aims to deconstruct the plurality of governance that has emerged. The second is a model, according to the *principles* of nodal governance advocated by Shearing and others, that seeks to address some of the negatives that have emerged from the nodal basis of the governance of security. This looks at addressing the inequities in security between nodes by, for example, empowering local communities to purchase security or by developing informal mechanisms of dispute resolution beyond the state (Johnston and Shearing, 2003).

It is not the purpose of this book to become involved in the expanding debates over the merits of this model of governance (for those interested in the debate see Loader and Walker, 2007; Wood and Dupont, 2006). However, the analytical framework used by the nodal governance 'school' provides ideal foundations for analysing security systems and the environments in which they operate. This section will therefore consider this framework and also seek to restructure some elements which are built on weak foundations and to complete some of the structures that have so far been neglected.

Central to understanding the nodal governance school is a return to Westphalia in 1648 and the emergence of the modern state alongside the writings of Thomas Hobbes (1651/1985) in *Leviathan*. Hobbes argued that without the state, society would descend into a war of all against all (state of nature) where life was 'nasty, brutish and short', and that it was in order to avoid this that men came together to make a social contract, building a state to provide security amongst other common goods. The cover of *Leviathan* provides a sound pictorial symbol of the social contract, with a friendly giant, made up of people, standing over a landscape, with a sceptre in the left hand signifying legitimacy and a sword in the right signalling coercive capacity. The use of one body made up of many signifies the aggregation of the many private interests of individuals into one common interest or commonweal (Shearing, 2006). These ideas of the state were given their modern form in the works of Weber, who distinguished the state as the 'human community that (successfully) claims the monopoly of the legitimate use of physical force within a given territory' (Weber, 1948: 78).

Changes in society, commerce and, most crucially, governance have begun to break down this traditional conception. The idea that the private realm is governed but does not govern has been challenged (Shearing, 2006). During the 1980s Shearing and Stenning published extensive research illustrating the importance of 'mass private property' as pockets of governance (see Shearing and Stenning, 1981, 1982). Others have sought to demonstrate the way in which the rise of transnational corporations has led to the emergence of firms that are now able to draw upon resources to rival many states (Strange, 1996) and whose systems of governance amount to 'private government' (Macauley, 1986). Research has also illustrated how the role of the state has changed, with one of the best analogies that of Osbourne and Gaebler (1993) who distinguish between the steering (directing) and rowing (doing) functions of the state, with the latter increasingly undertaken by the private and voluntary realms.

Nodes

These (and many other works) have led to the advocacy by some of the concept of nodal governance (for example, Kempa et al., 1999; Johnston and Shearing, 2003; Wood and Shearing, 2007). At this point it is worth analysing this perspective in more depth, and in particular, considering what nodes are. Central to the analysis is that the state is but one node amongst an assemblage of others. Shearing (2004: 6) has defined nodes as,

> ...locations of knowledge, capacity and resources that can be deployed to both authorise and provide governance. Nodes may or may not form governing assemblages, and they may or may not develop net-

works that traffic information and other goods to enhance their efficacy.

Burris (2004: 341) has developed this further, defining a node as a site of governance founded on four characteristics: mentalities, technologies, resources and institutional structure. It is worth quoting in depth his description of a node (2004: 341–2).

It need not be [a] formally constituted or legally recognised entity, but it must have sufficient stability and structure to enable the mobilisation of resources, mentalities and technologies over time. A street gang can be a node, as can a police station or even a particular shift of a firehouse. A node like this may be primarily a part of an integrated network, like a department in a firm; it may be linked to other nodes in multiple networks without having a primary network affiliation, like a small lobbying firm; or it may be what we call a 'superstructural node', which brings together representatives of different nodal organisations to concentrate the members' resources and technologies for a common purpose but without integrating the various networks – a trade association, for example.

A node could have a territorial basis, such as a town, a housing estate, a gated community, a shopping centre or a factory. However, a node need not necessarily be based upon physical space. It could be a community founded upon membership of a specific group, such as a church, a professional group or a sporting body whose membership is geographically spread. It could also be a community in cyberspace. The maturity and complexity of the governing structures within a node could vary from relatively simple informal arrangements to complex state-like structures. Nodes might be based in the state, for example, the police, or in the corporate, voluntary, informal or even criminal sectors. Nodes must also be distinguished from 'networks', which are much looser assemblages of agents, bodies and so on. Raab and Milward (2003: 417) define networks as '...a set of social entities linked directly or indirectly by various ties'. Though clearly – as with any definition in social science – there is a grey area in which it might be hard to determine whether an assemblage constituted a node or a network.

Nodes may fall within and/or cut across one another and an individual might be permanently located in some nodes, but also, in the course of a normal day, fall within the boundaries of others. Within this complex web of nodes, the governance of security is provided by an equally diverse range of agents. They range from state actors such as the police to security officers, staff with security responsibilities and volunteers. The security balance within each node is differently composed. The nodes also have constantly changing relationships with one another. A good way of illustrating nodal governance might be to illustrate how it affects me on a typical day.

I began my day by getting on my bicycle to ride to my office. I live in Southsea which falls within the nodes of Portsmouth (as a geographical/governance entity) and Portsmouth City Council. On the way to work, I stopped in the Southsea shopping precinct to go to the NatWest bank. This also falls within the Town Council boundary of Southsea. I arrived at work at St George's building, which is part of the University of Portsmouth. At lunch I went to the nearby shopping/leisure facility of Gunwharf Quays and ate a sandwich at Costa Coffee. I then returned to my office and in the late afternoon cycled home. Thus in that day I visited the following nodes: Portsmouth Police Basic Command Unit, Portsmouth City Council, Southsea Town Council, Southsea Precinct, NatWest, University of Portsmouth, St George's Building, Gunwharf Quays and Costa Coffee. At some points of the day I was in multiple nodes simultaneously, I was in other nodes for only a few minutes and in some the whole day. These all have varying degrees of governance over me while I am inside them, and each will use different technologies and have very different mentalities.

Completing the nodal structures

As noted above, there are a number of areas of the debate on nodal governance that require further 'construction'. One such area is that relating to the purpose and mentality of nodes. Most nodes have a lawful purpose and operate within the rule of law most of the time, but there are some that do not and others where there are lapses into deviance on a significant level or which operate within the rule of law, but that 'law' breaches fundamental universal human rights. Indeed Raab and Milward (2003), Wood (2006) and Wood and Shearing (2007) have raised the idea of 'dark networks' and 'dark nodes' operating illegally and/or covertly engaged in the delivery of governance. However, the association of the word 'dark' with deviance serves to perpetuate the many associations of this word and the related word 'black' with criminality, in turn, in my view, perpetuating the association of certain racial groups with crime and law-breaking. While there are some descriptions in common use that it would be difficult to change, such as 'black markets' – although even here many are now using the term 'illicit markets' – writers should not be creating new such associations. Therefore the proposal here is to use the term 'deviant nodes'. To develop this further it is necessary to recognise there are three ideal types of node, regular, quasi-deviant and deviant.

Regular nodes operate within the rule of law
Quasi-deviant nodes operate sometimes beyond the rule of law
Deviant nodes operate largely beyond the rule of law

Take for example the Provisional Irish Republican Army (PIRA) at the height of the troubles in Northern Ireland and to an extent today (in terms of policing crime). This was an organisation that undertook a campaign of violence with the aim of removing the British from Northern Ireland as well as acting as the de facto state in its heartland communities, 'solving' crimes and punishing the perpetrators. The terrorist campaign and the risks of 'normal policing' activity in Republican areas led to the Royal Ulster Constabulary (RUC) (now called the Police Service of Northern Ireland) focusing on 'counter terrorism', rather than on more routine crime, creating a gap which the PIRA were able to fill. Local residents were frequently afraid to contact the RUC and risk being labelled as informers, hence people preferred to contact the PIRA in situations where the police might more normally be the first port of call. The punishments carried out by the PIRA, ranging from warnings, severe beatings and 'kneecappings' to execution were also supported by many in the communities (Silke, 1998). The PIRA also represented a node of governance that extended beyond the physical space of its heartlands in Northern Ireland to its members worldwide. It also had systems of governance with defined punishments for their breach, security capacity to protect itself and the willingness and capability to administer punishments. However, it acted beyond the rule of law with activities and aims regarded as deviant by the British state and as such could be regarded as a 'deviant node'. Nonetheless, for many in its heartland areas it would not be the PIRA that would be regarded as a 'deviant node' but the British state. Such practices continue and also exist in the loyalist communities of Northern Ireland (Silke, 1998, 1999; Silke and Taylor, 2000).

Deviant nodes are found in many places. Extensive accounts exist of criminal organisations engaged in various levels of governance in the communities they live off. Hobsbawm (1959) shows how the Sicilian Mafia became a de facto state serving the interests of all to varying degrees. In a more specific study Chu (1996) describes how in Hong Kong, triads provide 'legitimate' policing services to the business community. More recent research on organised crime has also illustrated the networked and nodal nature of many groups (Wood and Shearing, 2007: 67).

It would be wrong to assume that 'regular nodes' never engage in deviant acts, because there are many that do occasionally breach the law. If we were to categorise every lapse as a deviation, there would be virtually no 'regular nodes'. However, there are some nodes largely anchored in the 'regular node' domain, which have significant covert deviant agendas. As such they deserve a category in their own right. The spectacular collapse of companies such as Enron, WorldCom and the Maxwell empire, based upon corrupt agendas driven from the very top provide examples of these. Another example can be found in Brazil where the state police and private security companies have been implicated in repressive social control, including organised death squads, most notably in the slums of large cities. For example, in Sao Paulo during the last three months of 1999, 109 citizens were killed by the Militarized Police and 69 were dispatched by off-duty officers working part-time in private security. As Huggins (2000: 121) has argued,

> In Brazil today, death squads continue to operate, paralleled by a growing 'rent-a-cop' industry, with each social control entity directly 'serving' different population segments to the benefit of one over the other. The poor are treated as undifferentiated 'criminals' by death squads and 'rent-a-cops', and richer Brazilians are protected from the poor by their private security forces.

Security in 'deviant nodes' will invariably have functions that conflict with the rule of law. Their responsibility is to protect their node regardless of whether this results in breaking the law. For example, an organised criminal group threatened by a potential informer to the police will seek to protect its node by silencing (murdering) the informant. Security generally operates legitimately and through the rule of law when employed in 'regular nodes', but it is important to note that even in 'regular nodes' security's responsibilities are to serve the interests of the node, which might not always be in the interest of the public at large.

Denizens

Linked to their nodal analysis, Shearing and Wood also make a case for 'denizens' rather than citizens as the inhabitants of nodes. They argue that increasingly people operate across multiple governmental domains and that the traditional conception of citizens and their rights and responsibilities, which held in a situation where individuals were linked to a specific domain, has become much more tenuous. Therefore the term 'denizen', which has emerged from the literature on citizenship and immigration, has been advocated, referring to '...an affiliation to any sphere of governance and its associated rights and responsibilities' (Shearing and Wood, 2003: 408). Thus 'denizenship' might be permanent or temporary depending on whichever node an individual is occupying at a particular point in time. Thus a node might contain staff, visitors, customers and so on who move in and out of the node for different periods of time but who while they are there are subject to the governance of the node. Some of these denizens may also occasionally become malefactors.

Malefactors

Nodes promote both their own and the state's objectives (in varying degrees). Frequently these objectives are breached, by criminal acts, unlawful acts or simply acts that contravene the organisation's own rules or norms. It would be wrong to term all these 'wrongdoers' criminals or offenders, because many of the breaches or contraventions of the nodal rules do not fall into the criminal category. Therefore the term 'malefactors' is proposed to capture those who breach the norms of the node, that is to say, someone who does something bad, but not necessarily criminal. The Cambridge Online Dictionary (n.d.) defines 'malefactor' as, 'a person who does bad or illegal things'. The word captures the essence of the adversaries of the node, those with whom the security forces are engaged in confrontation. The 'battle' does not always centre around crime. Equally problematic are 'bad' or 'wrong' things as defined by the node.

It is worth illustrating the benefits of this term by considering some of the discounted alternatives. 'Offenders', for example, might have been used, but it implies breaking the criminal law, and much of the deviance that security services are seeking to counter is not related to the criminal law, but concerns enforcement of the rules of the node or other civil matters, such as removing trespassers. Thus, 'offenders' and related terms such as 'criminals', 'felons', 'law breaker' and so on were dismissed. 'Combatants' was also considered, but although it is defined by the *Concise OED* simply as someone who fights, the tendency to use it in militaristic contexts evokes images of gun-battles and terrorists held at Guantanamo Bay and led to its being discounted. 'Malefactor' has also been used in computer security to cover those breaching security systems

(Kontenko, n.d.). Consequently the best term to describe the adversaries of the node was considered to be malefactors.

Malefactors can come from within or beyond the node, they might be denizens visiting the node or working there or they may even emanate from security as opportunistic malefactors or organised criminals planted to undertake an 'inside job'. Opportunistic malefactors stumble upon an opportunity to commit an act of deviance and do so. More determined malefactors might originate from any grouping, from organised criminal gangs to terrorist organizations and all points between. Malefactors can also be denizens who simply trespass or break a rule of the node by mistake. The discussion of malefactors will be developed further in Chapter 3.

The security system

The last major component of a node is the 'security system', that is, the social and technological elements designed to achieve security. The nature of the security system will determine the level of opportunities and the potential for malefaction (Tilley, 2005a). The function of the 'security system', or what Johnston and Shearing (2003) call 'security pro-grammes', is to guarantee peace (or security). They argue that it is composed of six elements. First, there is a 'definition of order', which comprises the norms and expectations underpinning a secure environment. Second, there is some element of 'authority' or 'authorities', which sets out the basis and justification for the 'definition of order'. Third and fourth are the 'technologies' used for achieving security and the 'mentalities' that determine the ways in which issues are constructed and addressed. 'Technologies' require resources and an institutional base so 'institutions' are the fifth element. Sixth is 'practice', which is the culmination of these elements. All bar technologies can be considered as part of a social system based upon human interactions and decisions.

At the nodal level the system of security can be composed of a diverse range of state, private, voluntary, human and technological elements. Ordinary denizens might have a security responsibility or specialised staff might be trained to undertake security functions. Trained staff are likely to be security officers provided by a commercial security company. There might also be a contribution to security from other nodes, most notably from the state in the form of the police. A good way of illustrating this is to look at a particular security system. Table 1.1 shows the main elements of the security system at Pleasure Southquay (Button, 2007a).

At Pleasure Southquay a managerial team composed of a generic operations manager (who was also responsible for services such as cleaning),

Table 1.1 The security system at Pleasure Southquay

Nodal context	Security system
33 acres of former MoD land	1 operations manager
65 shopping outlets; 20 restaurants, bars and coffee shops; 11-screen cinema; bowling complex.	1 contract manager
	1 security supervisor
	31 security officers
1450 parking spaces	Occasional extra officers
100,000 square feet of office space	Additional security officers and door
310 homes	supervisors employed in shops/bars
Further expansion of all of above	Ad hoc police presence on demand and
5000 + visitors at peak times	at high risk times
116 incidents of crime recorded in	Design, image and reputation
first year of opening	Limited entrances/exits linked to access control policy
	CCTV
	Rules

a security supervisor and a contract manager (who were both from the security contractor and were not permanently based at the centre although they spent a large portion of their time there) oversaw the functions of 31 security officers. For special events the 31 officers would be supplemented with officers drawn from other contracts. Many of the shops, bars and clubs also had their own security teams, smaller nodes in their own right. The police would also be there for ad hoc events and at times of high risk, such as late Friday and Saturday nights. It is also important to note the design, marketing strategies, access control measures and extensive CCTV system in the centre. Finally there were rules, though not in an overt sense at this node. All over the world similar nodes exist with such a security system. There are many other nodes in the corporate sector that have a dominance of private security. There are others dominated by a 'thin' state police presence, such as some housing estates. A 'thick' state presence can also be found at some state nodes where there are important people or where vital resources are kept, such as the Palace of Westminster or a military base. There are also some nodes that pursue more social crime prevention strategies, something that will be developed further in Chapter 7.

The success (or failure) of a security system – which will be the focus of Chapter 2 – is linked to the efficient (or inefficient) working of both social and technical elements. These tend to be mutually reliant. For example, an intruder alarm will only work effectively if it is the right tool in the right place, if it is installed correctly, maintained regularly, switched on and responded to when activated. All these actions rely upon human intervention. Thus it is important to stress that any security system is a 'socio-technical' system.

The *raison d'être* for the 'security system' for most nodes and individuals (individuals is used here to describe wealthy individuals who utilise security to protect themselves) is protection from something which is considered bad, most commonly some form of crime, but such events could also include acts of God, accidents, civil disputes and so on. It is important to note that the order that nodes promote in order to rationalise the need for protection is a particular order peculiar to that node, which in the case of 'deviant nodes' may also be an 'illegal order'. As Wood and Shearing (2007: 24) write in relation to large nodes such as Disney and General Motors,

> The order maintained ... is not simply designed to comply with the governance objectives of states. These corporations have their own objectives that they attempt to put into effect through a variety of governing strategies. In doing so they take into account the objectives of state agencies and may seek to enrol these in assisting them in corporate governance if this seems sensible.

For example, at the Sydney Olympics in 2000 security officers searched for and confiscated not only knives, weapons, bombs and so on but also disallowed the ostentatious display of brands that were not official sponsors (Chaundhary, 2000). Their role was focused upon a particular corporate interest as well as on the public interest. In another example a *Guardian* exposé of poor working conditions in some factories in India making garments for UK retailers disclosed the actions of security guards searching toilets to find slacking workers and force them back to their toil (*Guardian*, 3 September 2007). These are not crimes or even examples of anti-social behaviour, but they are behaviours defined by the node as requiring the attention of the 'security system'.

Before we consider some of the 'bad' things or risks (as they are more commonly known) with which the machinery of security must engage, it will be useful to explore the areas that they protect within a node. George and Button (2000) argue that security is used to protect four areas. The first area involves the *protection of people* or the denizens of the node. These may be the employees or customers of an organisation, residents, visitors or the general public. The second area relates to the *physical assets* of an individual or node, such as property, vehicles, capital equipment and the like. In the corporate sector the protection of the *information* of a node is often important. Many establishments have sensitive information, such as product specifications, lists of clients, financial plans and marketing strategies that could be very useful to a competitor or third party. The final area of protection, which again is largely in the corporate sector (though not exclusively) is the *reputation* of an individual or node. Reputation can be very important and the emergence of a negative image could result in serious damage to the individual or node. For example, when Gerald Ratner, then Chief Executive of Ratners, in a speech to the Institute of Directors in 1991 described one of the products in his jewellery chain as 'total crap' the resultant publicity wiped £500 million off the value of the company and contributed to its eventual collapse (*Telegraph*, 2007). Negative publicity, whether it is the result of a mistake, blackmail or genuine accident, can have a huge impact on sales.[1] Therefore, security is used, amongst other strategies, to protect people, physical assets, information and the reputation of an individual or organisation against the threat from malefactors.

As we have seen above, there is a wide range of sources of risk to the node that it is the function of the 'security system' to protect against. The following list outlines some of the most common.

1. *Protection from crime.* Depending upon the nodal context there are a wide range of potential crimes from which the security system will seek to protect the node. These include theft, burglary, fraud, robbery, assault, murder, terrorism and so on.
2. *Protection from protests.* Many organisations are targeted by protesters, which has the potential for disruption. For many nodes this is a key area of protection for the security system.
3. *Protection from deviant acts.* There are many forms of behaviour that are defined as deviant by nodes, but which aren't crimes. These might include parking where parking is restricted, or using a skateboard in a specific area, chewing gum or smoking.
4. *Protection from accidents and acts of God.* Frequently the security system will also have responsibilities relating to preventing accidents, fires and so on.
5. *Protection from non-denizens.* The final significant area of protection is at the borders of the node. For many, a key role of security will be to ensure that only the appropriate denizens gain entrance, whether that is a ticket holder for a concert, an employee with a pass or a person that 'fits' the image of what is desired by the node.

Conclusion

This chapter has explored some of the theoretical and conceptual foundations that will underpin the rest of the book. It began by examining what is meant by security and the emerging pluralisation of bodies delivering it. The remarkable expansion of private security, as one of the key elements in the delivery of security, was then briefly explored. The reaction to this expansion was then assessed identifying four main perspectives, the latter of which, the 'nodal positive', was examined in some depth. In doing so some of the key emerging themes in the literature of the nodal governance school were analysed. In undertaking this task the chapter then sought to develop some of the gaps in this literature and utilise the key parts for development throughout the rest of the book. In Part II, 'Security Undone', chapters 2 to 5 will explore some of the factors that contribute to undermining effective security. In doing so the chapters will examine the nature and extent of security failure, malefactors' views of security, the quality of the human element of security and the unsound foundations upon which private security operates.

Note

1   For example, the discovery in 1990 that Perrier water contained traces of Benzene led to the recall of 160 million bottles and a reduction in sales that some years later still hadn't surpassed pre-Benzene incident levels (Brand Failures and Lessons Learned, 2006).

Part II
Security Undone

2
Security Failure and the Security Myth

Introduction

Security failure encompasses a wide range of potential incidents that take place in different nodes and the consequences of which vary greatly. Failures of security in aviation resulted in the catastrophic events of 11 September 2001, while security failure in a supermarket might mean no more than the loss of £50 worth of razors. Another consequence might be nothing more dangerous than huge embarrassment and bad publicity, such as in the incident in which a member of Fathers 4 Justice breached security at Buckingham Palace to reach the Queen's balcony in a Batman outfit in order to highlight the organization's campaign for greater access for separated fathers to their children. At another extreme, security failure might result in the theft of goods worth millions of pounds, as in the Brinks Mat robbery of 1983, where £26 million of gold bullion was stolen. Security failure occurs all the time despite the millions of pounds spent to reinforce security systems. As Zedner (2003a: 158) argues,

> ...absolute security ... is a chimera, perpetually beyond reach. Even if security were today obtainable ... the potentiality for new threats means that the pursuit can never be said to be over ... Just as the capabilities and intentions of potential adversaries are unknowable, so there may be unknown vulnerabilities, revealed only when they are exploited. The central issue here is that security is not and can never be an absolute state. Rather it is a relational concept whose invisibility must be continually tested against threats as yet unknown.

Indeed in many nodal contexts economics dictate that degrees of security failure are actually tolerated. The approach of many retailers to shoplifting exemplifies this. Shops want to maximise sales, which is incompatible with tight security, so a degree of security failure is often tolerated, with the costs passed on to the consumer. There is of course a level beyond which it becomes necessary to pursue greater security and there are some nodal contexts where absolute security is an aspiration – such as nuclear power stations, banks holding large sums of money or the residence of the prime minister or president of a country. Zedner is, however, over-pessimistic about the potential of security. As this, and subsequent chapters, will seek to demonstrate there is much that can be done to improve the effectiveness of security. This is not to deny the mythical status of absolute security, which Zedner rightly states is not possible, but to propose the creation of security systems that reduce the risk of security failure to an absolute minimum.

This chapter begins its analysis with a more detailed examination of security failure and with an exploration of the extent of security failure. It will then consider some of the literature on accidents and disasters and highlight how the application of some of the theories here can be used to explain security failures. Three detailed case studies – the Gardner Museum Heist in Boston in 1990, the attacks of 11 September 2001 and the breach of security at Prince William's birthday party in June 2003 –serve to highlight the causes of security failure and show how disaster theory can be applied. The chapter concludes with a model of security failure.

Before we embark upon this, however, it is worth noting the following caveats. Security failure is not an easy subject to study. First, the embarrassment of security failure can be such that the breach is never actually publicised (many frauds for example). Second, in order to minimise the chances of similar breaches happening again, detailed information of what went wrong is often not made publicly available. Consequently, the study of security failures is restricted to cases where information is available, which may well be a biased sample in the first place. Nevertheless, particularly in the public sector, when there is a major breach there is often some form of enquiry (though not always fully published) and information can be gleaned from these (Woodcock, 1994; Learmont, 1995). Sometimes the media carry articles that shed light on security failures, both reports from journalists and, in some cases, from the perpetrators themselves some years later (which phenomenon will be developed in Chapter 3). Finally there are also evaluations of particular security products, which can shed light on the causes of security failure and which can be used to further our understanding (see Gill and Spriggs, 2005 for example). The examples of security failure discussed in this chapter must, therefore, be treated as a group selected by the author on the basis that there is enough information available to illustrate that there was a security failure and, in some cases, enough to show why security failed.

Security failure

First, we should define what we mean by 'security failure'. Security failure enables an act that breaches what the security system is designed to prevent. This could be a criminal act, such as a robbery, burglary, theft and so on; or a lesser act, such as trespass or breach of organisational rules. Security failure may also encompass the enabling of 'good acts' against 'deviant nodes', such as an attempt by the police to infiltrate a criminal gang. However, for the purposes of this book the focus will be on security failure in 'regular nodes'.

Failures in security are taking place all the time. By the very nature of security failure it is difficult to gauge its extent, although in some sectors there are statistics that provide a barometer of the success or failure of security in certain contexts. This section will trawl some prominent sectors to provide some indicators on the extent of security failure. It will also provide some anecdotal illustrations of security failure where there is less statistical evidence.

Aviation security

In August 2002 a man was arrested trying to carry a gun on board a flight bound for London from Stockholm. The gun had been discovered by a security guard undertaking normal security screening procedures at the airport (BBC News, 2002a). This is how we would like to imagine that airport security works. Unfortunately there are many more exposés and other evidence to illustrate that aviation security fails all too often (Hainmüller and Lemnitzer, 2003). Aviation security is one of the most salient and important examples of security with which ordinary members of the public come into contact. Despite having one of the highest profiles and notably strict regulatory frameworks, there is a great deal of evidence – both anecdotal and, more significantly, statistical – exposing fundamental weaknesses in aviation security (something which is explored further below in the case study of September 11th).

Airports are generally divided into two areas and to go 'airside' passengers have to be screened or have appropriate credentials to pass. The protection of airside from intruders and the screening of passengers, luggage and cargo are significant responsibilities (George and Button, 2000). The ultimate aim is to ensure that unauthorised items (and deviant persons) do not find their way on to the plane. For passengers, searches focus on hand weapons, explosives and substances that could be turned into weapons for the purpose of hijacking the plane or seeking to blow it up. For the luggage and cargo, the focus has been on finding explosives. Underpinning these security regimes is an international framework of agreements that are enacted into regulatory measures at state level and enforced by state agencies. In the UK the relevant agency is TRANSEC (Transport Security Division) a section of the Department for Transport, in the USA it is currently the Transport Security Administration (TSA), which is a part of the Department for Homeland Security (Wilkinson, 2006). Among the strategies used by agencies in countries responsible for regulating security is the testing of security by sending undercover inspectors through security screening with weapons and explosives. In the USA results of these tests and other relevant statistics have been publicly available in the past and it is possible to examine these.

In 1978 US screeners failed to detect 13 per cent of unauthorised objects (United States General Accounting Office, 2001). By 1987 the performance of security had worsened, with 20 per cent of those attempting to take weapons through security avoiding detection. Indeed, during the first two sweeps the FAA imposed $2 million of fines on security operators (Wallis, 1993). Data from 1991 to 1999 showed performance declining further, but by this time the data were designated as 'sensitive' and were not released, although the performance of security screeners was thought to be of similar if not worse levels. Indeed a recent Congressional Report noted:

> More recent results have shown that as testing gets more realistic – that is, as tests more closely approximate how a terrorist might attempt to penetrate a checkpoint – screeners performance declines significantly.
>
> (United States General Accounting Office, 2001: 2)

Post 9/11 there have been significant changes in screening, with the replacement of contract staff with an in-house force of security screeners, but reports have continued to raise concern over the level of unauthorised items getting past security on tests, even if there are now no publicly available statistics (Strohm, 2004).

On access to secure areas official reports have been equally critical of US airport security. In one example it was found that inspectors could gain access to secure areas as unauthorised persons on 75 per cent of all occasions (US Department of Transport, 1993). During 1998–99 further tests on gaining access to security areas were conducted and inspectors were able to gain access to secure areas 68 per cent of the time and to board aircraft on 117 occasions. Following release of the report further tests were conducted between December 1999 and March 2000 and access was still gained in 30 per cent of attempts (United States General Accounting Office, 2001).

Post 9/11 aviation security has become a much more salient issue, with measures introduced in most countries to further enhance security (Hainmüller and Lemnitzer, 2003). Unfortunately statistics are no longer publicly available in the USA (and have never been in many other countries). Nevertheless there is ample anecdotal evidence illustrating the inherent weaknesses in aviation security, more often than not exposed by journalists. Just a few months after the 9/11 attacks, Richard Reid was able to board a flight from Paris to Miami with explosives hidden in his shoes. In September 2002, one year after the September 11th attacks, in the week approaching the anniversary, two separate journalists managed to breach security in the UK. One managed to get onto a flight at Heathrow carrying a meat cleaver and a dagger, while the other carried a replica pistol (BBC News, 2002b). In another exposé in 2004 at Humberside airport two journalists were able to walk through a gate and stroll unaccompanied around the aircraft in a manoeuvring area for over half an hour (BBC, 2004). Similarly at Heathrow during the same year a reporter was able to breach Heathrow security, wandering without challenge through secure areas in British Airways offices – even finding a security manual, walking up to passengers waiting to board a plane on the tarmac and on another occasion walking around an aircraft (The Sun Online, 2004). However, possibly the most disturbing exposé was by an undercover reporter at Birmingham (UK) airport in 2007 in a programme showing shocking lapses in security, which included security officers sleeping on duty, claiming to take drugs while on duty and most alarmingly, an officer reading the paper while supposed to be watching an X-ray screening machine (BBC News, 2007a).

The above are just a few examples from the UK, but a Google search reveals regular examples from throughout the world. Journalists provide a good comparison to a terrorist as outsiders seeking to breach security. Many of the reasons for the failure of security at airports derive from the inability of security staff adequately to undertake their duties of search. Some of this relates to competence issues, but there are also more systemic causes, such as recruitment strategies, numbers of staff available and the organisation of their frequently monotonous duties. Some of these issues will be developed further in Chapter 4. Nevertheless one is left with the conclusion that aviation security measures are not always adequate in either the UK or the USA and that a determined terrorist would have a reasonable chance of breaching security and smuggling either a weapon or a small explosive onto a plane.

Materialistic crimes: robbery, theft, burglary, fraud and so on

The most common security failures relate to materialistic crimes where malefactors carry out a robbery, theft, burglary, fraud or the like for material gain. One could consider each crime as a security failure, but this would make the concept so wide as to be useless, particularly in terms of crimes against ordinary people. For instance if a car is vandalised on the street it would tell us little about security failure, because there are unlikely to be any security measures in place to protect a person's car from such a risk. For that one would have to rely on police patrolling the area, CCTV coverage and so on. There are of course cases where it is possible to assess security failure in a household, such as in the case of a burglary. Were windows left open? Did the intruder alarm fail? Was the back door left unlocked? And so on.

However, the focus here is on commercial and public organisations. They often have extensive and sometimes quite complex and expensive security systems in place to protect themselves from a wide range of potential risks, and their experience of security failure consequently becomes more useful for an analysis. Take for instance shoplifting against retailers, a crime that is never going to be completely eliminated, but which can be reduced, and in relation to which individual retailers might demonstrate different degrees of success in combating the problem or, to put it another way, different levels of security failure. Assessing the different security systems in place will shed light on success and failure. Many of these malevolent actions, such as shoplifting and criminal damage, are never going to be completely stopped, rather the aim will be to reduce their incidence to a manageable residual level. Unfortunately comparative data are very rare and it is therefore difficult to compare the levels of success of different organisations, but before some of the few studies that have been published are considered, we will look at data that estimate the size of the problem.

The Home Office has extrapolated from data in the UK crime statistics to estimate crimes against the commercial and public sector for 1999–2000 (Home Office, 2000: 13). It found there were about 70,000 robberies, 960,000 burglaries, 29 million thefts from shops, 40,000 thefts of commercial vehicles, 60,000 thefts from commercial vehicles, 270,000 thefts by employees, 1.4 million thefts by others, 3 million acts of criminal damage and 9.2 million cases of fraud or forgery.[1] The same report also sought to estimate the costs of such crime, including the actual losses and the costs of security measures, as well as the costs to society at large. It found that a burglary cost £2,700, theft from a shop cost £100, theft of a commercial vehicle cost £9,700, theft from a commercial vehicle cost £700, criminal damage cost £890, and robbery or a till snatch cost £5,000 (Home Office, 2000: 45). These statistics illustrate the significant number of security failures occurring in commercial and public sector organisations and their substantial costs.

In the UK the British Retail Consortium (BRC) also conducts regular surveys of the extent of crime against retailers. In 2005 it found that the average annual losses over the past five years were £2.24 billion (British Retail Consortium, 2005). Subtracting from this the £700 million costs of security, the near £1.5 billion left could be read as the cost of security failure. Depending on the interpretation of the data it could be stated that despite £700 million invested by retailers in security there was still £1.5 billions worth of failure or, alternatively, that the investment in security is the reason why failures are held at that level and are not much more costly. In the USA the National Retail Federation conducts a survey relating to 'organised retail crime' which is defined as, 'theft of merchandize where multiple people working together steal large quantities of goods from retail stores' (National Retail Federation, 2007: 1). In a survey that was claimed to cover all retail sectors in 2007, 79 per cent of respondents had been a victim, with 71 per cent noting an increase over the last 12 months. These examples from retailing give some illustration of the costs and extent of security failure to retailing (although some retailers might argue that this represents a degree of success, since if it were not for their interventions these figures would be even higher).

Perhaps a better way to illustrate security failure in this arena is to highlight some of the most spectacular heists from across the globe noting the extent of the heist and from the evidence available what the principle security failure was. One of the biggest robberies in British history was the 1983 Brinks Mat robbery in which gold bullion valued at £26 million was stolen from a warehouse at Heathrow airport. The robbers, dressed as security guards, were able to gain entrance and once inside the warehouse threatened the security guards, using firearms and by dousing some in petrol, forcing the guards to reveal the appropriate security codes. This enabled access to the bullion which the robbers were then able to remove from the warehouse. The subsequent trial revealed that one of the Brinks Mat guards, Tony Black, was the brother-in-law of the gang leader and had passed on detailed information regarding security procedures and access through the main door (Hogg et al., 1988). Security failure in this case, therefore, was principally down to the corruption of one security guard. However, could more extensive vetting of guards have revealed a potential risk? Should more extensive security procedures have been put in place when such large amounts of bullion were being held? It is therefore questionable whether it is appropriate to blame this security failure on one 'bad apple'. Systems should surely always be designed to minimise the potential consequences of the corrupt behaviour of one junior member of staff.

The world's biggest diamond heist took place in Belgium in February 2003 when over €100 million worth of diamonds were stolen. Antwerp is the world's diamond cutting capital and the raid took place in the Diamond Centre, in the main vault, where 123 of the 160 safety deposit boxes were opened (BBC News, 2004a). The Centre is heavily protected, with barriers restricting vehicle entrance, regular police patrols, 24-hour security guards, CCTV cameras, alarms, security bars, vaults and so on. The raid took place during the weekend of 15–16th February when the city was distracted by the Diamond Games tennis tournament. Without using any violence against security staff the gang were able to breach extensive security systems. At 7 p.m. on Friday the vault doors closed and at midnight the following evening at least three men took the elevator to the vault. Here the team had already disabled the motion detector, sprayed the area with silicone and taped over a light detector, giving the gang freedom of movement. They then broke into an adjacent room where they knew that a key to the vault was held and used this with a security code (the police do not know how this was gained) to secure access. They then assembled a tool that enabled them to open each safety deposit box and once fully loaded they used copied keys to open the doors to leave the building. Before doing so they went to the CCTV room to replace the video tapes with the previous night's footage (CourtTV.com, n.d.).

The raid was rumoured to have been undertaken by a highly professional gang of Italian criminals brilliant at defeating security systems.

The diamonds have never been recovered and the only prosecution in this case was of an Italian, Leonardo Notarbartolo, who had worked as a diamond merchant at the Centre some years earlier. He was charged with theft, use of false documents and possession of false keys, but not in relation to the heist itself (BBC News, 2004a). The implication of the Notarbartolo prosecution was that merchants had knowledge and access which in the hands of one corrupt person could be used to mount a heist. Clearly the thieves who undertook the raid were also incredibly professional. In the absence of detailed information on the raid, it is only possible to speculate that there was a flaw in the design of the security system overall in that too many people were familiar with it, and it was possible for one corrupt person to be able to exploit this.

Protests and stunts

Protesters and others frequently resort to stunts with the aim of maximising publicity for their cause from the exposure the incident receives. Some groups, among them in the UK Greenpeace and Fathers 4 Justice, have built a reputation on breaching security at high profile locations such as the Houses of Parliament and Buckingham Palace. The security consequences of being unable to prevent such stunts are not as significant as failure to prevent terrorist attacks, the results are only red faces and bad publicity. Nevertheless they do shed light on weaknesses in security. Again there are no statistics available on the number of these types of breaches, so anecdotal examples will be used.

The Palace of Westminster has extensive security measures that are intended – at one extreme – to protect it from terrorist attack, and at the other to ensure the enforcement of Westminster rules, such as what bars a 'stranger' (non-member) may drink in. Security resembles that found at an airport, with screening of all non-passholders seeking entrance. Passes give admission to different areas. On the 19 May 2004 during Prime Minister's Questions, Tony Blair was hit by purple flour thrown by a Fathers 4 Justice protester sitting in the public gallery. Luckily this was a stunt and the missile was only flour, but it could have been Ricin or Anthrax, and the breach sparked an MI5 enquiry into House of Commons security (BBC News, 2004b). A screen has now been placed in front of the public gallery. Screening procedures are designed to detect weapons and a small amount of powder is virtually undetectable unless a full body search is undertaken, something not feasible in the Palace of Westminster.

A few months after this incident, security at the House of Commons was breached yet again when pro-hunt protesters[2] stormed the Chamber of the House of Commons while it was in full session. It was revealed by the Speaker that the eight protesters had gained access to the Palace by claiming to be attending a Committee meeting. He stated that,

> Once there, they were led into the small stairway to the north end of the corridor – probably by a passholder who was clearly exceeding his or her authority. It was not clear whether the passholder who apparently helped the intruders was an MP, a reporter or an employee of a member.
>
> (BBC News, 2004c)

The report by the BBC revealed that the protesters had undertaken extensive planning and had even made a dry run. They had clearly secured 'inside' knowledge on the strengths and weaknesses of parliamentary security. In this case the security breach was probably largely down to the 'corrupt' behaviour of one or more passholders at the Palace of Westminster. However, it raises the question of whether the security system is adequate if all that is required to breach security is one corrupt passholder out of the many thousands who work in the House of Commons.

But it is not just politicians who can become the target of protesters or others bent on performing some kind of stunt. On 22 February 2001 in Germany the pop singer, Robbie Williams, while in mid-song, was pushed off the stage by a man who had secured access from a backstage area. The song was stopped while Robbie got back on stage and the intruder was dealt with. Luckily Robbie did not suffer any physical harm as a result of the incident. It transpired that the intruder had psychiatric problems and was motivated by a belief that the person on stage was a 'clone' whom he wished to expose (BBC News, 2001a). It further transpired in later interviews that the intruder had gained access to the stage simply by running past a member of the security staff who was protecting the stage (LiveLeak, 2007). The system might have prevented this incident if there had been more checkpoints to go through to reach the stage, more security staff or additional technology to protect the stage. Thus, this is another example where the security system had not been designed effectively enough to prevent a breach occurring, and where the last resort, the human element, had also failed.

Violence

The final area that will be considered in terms of security failure is that of specific acts of violence. Violence is unfortunately all too common in most societies and while many incidents involve friends, family members, partners and so on and occur spontaneously where there is no security system in place, there are also situations in which violence occurs in spite of the fact that security systems are in place. Many politicians and celebrities are at risk of attack and as a consequence have security measures to protect them. There are some groups of workers who regularly face violent people, such as doctors and nurses in accident and emergency departments of hospitals (Waddington et al., 2005). Finally there are some groups that are perceived as much more vulnerable to attack, such as schoolchildren and students, and for whom special measures are put in place.

The International Crime Victimisation survey shows that in Western Europe 3.6 per cent of men and women have experienced assault at work, with 7 per cent of women experiencing a sexual incident. This compares to North America where 2.5 per cent of men and 4.6 per cent of women had experienced assault, with 7.6 per cent of women experiencing a sexual incident. In England and Wales the survey found 3.2 per cent of men and 6.3 per cent of women had been assaulted, with 8.6 per cent of women experiencing a sexual incident (Licu and Fisher, 2006). The costs of this problem are illustrated by the fact that in the USA it has been estimated that between 3 and 7 million working days are lost each year as a result of workplace violence. Below we look more closely at two areas where violence has become a major issue: healthcare and education.

In the UK violence against National Health Service staff has become an increasing political priority and measures to tackle the problem have been introduced. The 2005–06 statistics for the NHS showed that there were 58,695 incidents of assault (NHSCFSMS, 2006). A typical example was reported in February 2007 when a drunk driver who had been involved in an accident was convicted of assaulting the nurse who had been trying to treat his injuries. The man struggled, shouted and swore at staff, which meant that it took two and half hours to deal with his injuries rather than half an hour. He was fined £700 for the assault in addition to a further £650 fine and an 18 month driving ban for the drink-driving offences (BBC News, 2007b). Some healthcare workers have experienced much more serious attacks. For example, in 2004 a William Kerr was given a life sentence after being convicted of assault and intent to ravish after perpetrating a brutal attack on a nurse on the nightshift of Penrith Infirmary. He threatened the nurse with a knife, punched her and tore her underwear off in what was described as a 'terrifying sustained attack'. Kerr was told he would serve a minimum of 10 years and it transpired he had two previous convictions for sexual assault (BBC News, 2004d).

From the limited information available on these two attacks it is difficult to determine if they could have been prevented with better security, although one suspects that a well-trained security presence might have helped. However, there is an example from the USA which does show how security failed. In 2004 Dr Erlinda Ursua was murdered at the John George Psychiatric Pavilion, California by one of her own patients, a woman with severe mental illness. The doctor was bludgeoned to death in a treatment room, while outside other staff went about their business. Erlinda's body was discovered by a cleaner later in the day and it was reported that the police believed she had lived for about an hour and a half after the attack. To many staff the attack did not come as a shock, as assaults were common and the police frequently delivered psychiatric patients to the unit without having taken weapons from them. Despite the existing problems the hospital management had done nothing although a fine of $30,000 had been levied six months before the murder for the failure to report two attacks on staff, as required by state law. A simple security procedure that could have prevented this murder would have been a requirement for a third person to be present. Indeed the inspection that led to the $30,000 fine had recommended the introduction of a mandatory 'buddy' system, alongside CCTV surveillance and police patrols of the hospital. The board response to these recommendations was to pledge to improve security, but no comment was made on the specific proposals advocated in the inspection. Workers at the facility claimed that little was done and even basic recording of incidents still seemed to be problematic. The security staff who worked there, were alleged to be 'scared' and incapable of dealing with incidents. Indeed in one incident, where a nurse was been stabbed, a security guard who was nearby refused to intervene claiming he didn't get paid enough for that (Weitz and Luxenberg, 2004). This case shows systemic failures within a culture of toleration of violence against staff, with little action taken and generally poor security. As will be shown later in this chapter there are many examples, like this one, where the organisation could have learnt from its own experience and worked to prevent such incidents occurring. Simple procedures could have been implemented to prevent this particular crime.

School, college and university campuses have also become common sites of serious violent incidents. Columbine and Virginia Tech in the USA and Dunblane in the UK are notorious for the massacres that bear their names. However, another incident on a US campus provides an example of a more common type of security failure that had the most severe consequences. On 5 April 1986 at Lehigh University, USA, Jeanne Clery was raped, sodomized and murdered in her own room. The murderer was a drug and alcohol abuser who gained access by walking through three propped-open doors. The incident was portrayed by the university as a tragic one-off incident. Research by the family, however, revealed that there had been numerous violent incidents on campus and 181 reports of propped-open doors in her dormitory in the previous four months. They also discovered that violence on campuses in the USA was a widespread problem. After successfully pursuing a legal action against the university the family established a campaign for security on campuses which culminated in the Crime Awareness and Campus Security Act 1990 which requires all colleges receiving Federal funds to make prospective and existing students aware of their crime statistics (Security on Campus, 2001). Clearly this security failure occurred primarily because doors were left propped open, but it is also important to examine what systems existed to ensure they were closed and if they did exist why they were not pursued?

Learning from disasters

In seeking to develop a broader theoretical perspective of why security fails it is worth examining the extensive research on organisational failure and disasters. Before we can consider some of the incidents above in relation to accident theory, however, a fundamental question arises. Are security failures and disasters comparable, given that one is based on human malevolence and the other is usually the result of a combination of many factors that in most cases do not involve purposeful deviant human behaviour? To answer this we should note that accidents/ disasters are commonly divided into three groups: 'natural', 'technical' and 'social', the latter two of which are 'man-made' (Borodzicz, 2005). 'Man-made' disasters can be compared to security failure. Table 2.1 offers a comparison.

Malefactors should therefore be viewed as any other potential 'pressure' likely to cause a hazard (see Table 2.1). As with many accidents and disasters security failure occurs because of a failure in the socio-technical system. Following Cohen and Felson (1979), for security

Table 2.1   Comparing pressures on a dam and a safe deposit centre

A dam	A safe deposit centre
Water pressuring dam with natural desire to burst through if possible.	Malefactors lurking with desire to steal valuables from safe deposit centre.
Socio-technical system in place through design, science, tests, risks, maintenance etc. to prevent dam collapsing.	Socio-technical system in place through design, science, tests, risks, maintenance security personnel etc. to prevent safe deposit centre being attacked.

failure to occur, there needs to be a suitable victim, a lack of a capable guardian and most significantly a motivated offender. For an accident, to use the same analogy, all that is needed is a suitable victim, the absence of a capable safety system and a pressure or incident that exposes it. The advantage for a security system is that even if the system fails there might not be a malefactor present to exploit it; whereas with a safety system, disaster is usually inevitable, for example, a bridge that collapses as a result of stresses from too much traffic.

For some theorists accidents are largely the result of failures in organisational systems (Perrow, 1984). For Perrow 'normal accidents' are the result of systems malfunctioning, particularly where there is a reliance on tightly coupled systems. For example, a hotel burns down because the sprinkler system fails to work because of a faulty fire alarm. Clearly there are comparable examples in security failure where, for example, a burglary might occur because the burglar alarm failed to activate because of a faulty motion detector. However, perhaps more relevant are the works that link disasters to systemic failures in either the human or technical elements of the system or in the two elements combined (Borodzicz, 2005). Ultimately the success or failure of the system rests upon these two elements. As Cox and Tait (1991: 93) argue, 'The majority of accidents are, in some measure, attributable to human as well as procedural and technological failure.' Writers such as Turner (1978) argue that disasters are best understood as the result of a number of interrelated decisions, systems and consequences that build upon each other. It is therefore more useful to view disasters in terms of the technical and social problems that *interact* to produce such incidents. Turner's ideas can be seen in his disaster sequence model (DSM), illustrated in Table 2.2 below (Turner 1978).

Turner's model consists of three distinct but interrelated stages. In the first stage, before the 'disaster' occurs, the context in which the disaster

Table 2.2   Turner's disaster sequence model

Stage1
• The incubation period
• The operational social technical system
Stage 2
• The precipitating event
• The disaster itself
• Rescue and salvage
Stage 3
• Inquiry and report
• Feedback

will later take place is created with a precipitating decision, such as the decision to build a dam or bridge, to open a new department or to engage in a new course of action. Once this decision has been made, the environment for the disaster is 'incubated' and the foundations for subsequent events and actions begin to build up. A socio-technical system becomes operational and this is the environment in which procedures and decisions operate vis-à-vis humans exploiting that context. In the case of the building of a new dam, this would involve the construction phase followed by the operational and maintenance procedures.

The second stage of Turner's model includes the triggering event and the disaster itself. The triggering incident may be influenced by the events and actions of the previous stage with the build-up of latent failures in the system. The more failures latent within a system, the greater the likelihood of disaster. For example, a bridge which is regularly over-burdened with traffic in excess of its safe limits (the latent failure) might be hit by a major storm, leading to catastrophic stress and the collapse of the bridge. The triggering event is the storm, but the socio-technical system that enabled excessive traffic to use the bridge also affected the outcome. The final part of the model, then, encompasses the learning process after the event. Inquiries and feedback provide for organisational learning with the aim of reforming the socio-technical environment so that the disaster does not occur again.

These ideas could clearly be applied to security failure. Take for example the Jeanne Clery murder above. The security failure relates to enabling an intruder to get into her room and rape and murder her. However, this disaster can be traced to the decision to build the dormitory and establish a socio-technical system to secure it. The crimes that had previously occurred on campus, but which weren't publicized illustrated the potential risk. The propping open of doors provided the opportunity, but the reason they were open was because of the socio-technical system in place amongst students, caretakers and security staff which did not treat their closure as necessary. The consequent enquiries led to legislation requiring students to have access to information on crimes so that they could make informed decisions about the risks at a particular college and over risk avoidance or reduction. No doubt action was also taken after the event to end the practice of propping open doors as well as to install additional security measures on the campus. Like disasters, such tragic events also offer important learning opportunities for organizations.

Toft and Reynolds (1997) argue that disasters continue to occur for the *same or similar reasons* and that it is therefore possible to learn from these events and to use knowledge of how and why they occurred to reduce the likelihood of similar disasters happening in the future. This also applies to security failure. The history of disasters is littered with examples in which organisations failed to learn from their own and other organisations' mistakes. For example, the Kings Cross London Underground fire in 1987, which killed 31 people, was not the first example of a fire on a wooden escalator on the London Underground. There had been a serious fire at Oxford Circus in 1985 and a number of fires had also occurred at other stations. At a meeting organised after the Oxford Circus fire, the management of the London Underground resisted requests that they call the fire brigade to every future suspected fire. If they had done as requested, the consequences of the Kings Cross fire might not have been as severe (Toft and Reynolds, 1997).

Toft and Reynolds identify three types of learning that can be informed by a disaster. First, there is *organisational learning* where individuals within an organisation draw their own lessons from an event. An example might be where an individual in an organization learns that a colleague has lost all his computer data as the result of a virus from which his computer could have been protected by anti-virus software, which he hadn't bothered to install. The other person then decides to install the software to counter this risk.

Second, there is *isomorphic learning* in which a disaster (that occurred in another place, at an earlier time, or to another business or organization) is studied by other similar groups. The intention of scrutinizing 'someone else's disaster' is to identify and assess potential risks that might apply to your own systems or procedures and to eliminate potentially hazardous environments or practices in your own industry or organization.

Finally, there is *iconic learning* where simply being informed of a negative event is considered a learning event in itself. Thus, to use the virus example again, a PhD student is reported on the news to have lost the whole of his PhD thesis three weeks before he is due to submit as the result of a virus which could have been easily protected against with appropriate software. Neither had the student made a copy of his work. This report might encourage other students to make back-up copies and to install appropriate anti-virus software. These types of learning can be further divided between passive and active, between simply knowing about something and actually doing something about it. Toft and Reynolds argue that while the first and third types of learning are important, the most significant is isomorphic. The benefits of isomorphic learning will be explored further in Chapter 6. The three case studies below will be used to model the causes of security failure and to assess further the ideas of Turner, Toft and Reynolds.

The Gardner Museum Heist, Boston, USA

The biggest art theft that the world has ever seen took place on 18 March 1990 at the Isabella Stewart Gardner Museum in Boston, USA. Thieves escaped with 13 artworks, including paintings by Vermeer, Rembrandt, Degas and Manet, worth over $300 million. To date none of these paintings have been found (CNN.com, 2002). The heist began just after 1 a.m., when what appeared to be two police officers came to the side entrance of the museum and buzzed through on the video intercom to the two security guards on duty, claiming there was a 'disturbance on the grounds' that needed to be investigated and asking to be let in. The guard let them in, which was the first mistake. As Lyle Grindle, the Museum's head of security after the theft stated, 'The policy has always been that you don't open that door in the middle of the night for God. Why on this one night they opened the door no one can explain' (CNN.com, 2002). When the 'police officers' reached the security guard, he made his second mistake, allowing them to trick him into coming out from behind the desk, where the only alarm directly linked to the police was located. They did this by claiming he looked familiar, telling him that there was a warrant out for his arrest and demanding to see his identification. Once he was out, the thieves spreadeagled him against the wall and handcuffed him. Meanwhile the second guard returned from patrol and was also handcuffed. The guards were then bound with duct tape and put into the basement, positioned 100 feet apart. The thieves then went about their work and were out of the museum by 2.45, 84 minutes after they had arrived. Before leaving they removed the video tape from the CCTV recorder at the security desk and ripped the computer printout from the motion detector equipment (although the information was still recorded on the hard drive) (Boston.com News, 2005).

Before the reasons for the security failure are analysed it is worth examining the composition of the security system during the night. The human element consisted of two security guards and both were inexperienced. The guard who let the thieves in admitted to having occasionally been 'stoned' while on duty. He was a student by day, played in a band in the evenings and then arrived at the museum to do what he described as '...the most boring job in the world'. He also admitted that some months earlier, during the Christmas holidays, he had invited some friends into the museum to get drunk and admire the paintings and building, but he insisted that on the night of the heist he had been sober. The second guard normally worked days, but was brought in to cover the sickness of a more experienced guard, who had been working nights for several years. On normal nights the guards would take it in turns to patrol the building and watch the four video monitors. Throughout the museum motion detectors would send a silent alert to the security desk if activated. There was also one alarm linked to the police that could be activated from behind the security desk. Paintings carried warning bleepers that activated if they were moved, largely for daytime security. One of these was activated during the raid, but the thieves just smashed it (Boston.com News, 2005).

At first reading this security failure looks to be entirely down to the incompetence of the security guard who let the 'police' in. The security director of the museum claimed that it was the policy of the museum not to let any one in, including the police, unless their presence was requested. This policy was written in the manual kept in the security desk. The security guard claimed he did not know if the policy applied to the police. The risk of this failure occurring was exacerbated by having two inexperienced guards on duty neither of whom were clear on this policy. The second mistake, in which the guard came out from behind the security desk, illustrated a weakness that was recognised by the museum, and a report on security a year before had actually advocated that the security desk be moved into a control room only accessible by a key.

There are therefore a number of systemic failures at work in this security failure. First, there was a failure in the recruitment strategies of the museum. To entrust such a vast range of valuable treasures to guards to whom 'it's a short-term job', rather than to 'professionals' to whom security is a career was a fundamental error. This is not only a problem in terms of the motivation and commitment of the security staff, but also exposes intimate knowledge of the security system to a larger pool of people who may purposefully or by accident reveal this to malefactors. Second, to allow two inexperienced guards to be on duty alone is another failure. Third, there appeared to be a lack of effective supervision of the guards – given the admission that they were able to be 'stoned' while at work and to let friends into the museum. Training is the fourth failure, given the dispute over the policy on the entrance of the police. If training had been effective there would not have been a dispute. A fifth failure is the location of the security desk, alarm and video recorder. These should all have been in a secure location. Finally, given the value of the artefacts in the museum, two layers of security would seem inadequate. The more layers of security there are to beat, the more mistakes a security system can tolerate. Thus although on the face of it this seems a simple case of human failure an analysis reveals a series of systemic failures at work, creating the conditions for this human failure to take place.

11 September 2001

On 11 September 2001 the worst terrorist attacks that the world has ever seen occurred when four airliners were hijacked and turned into suicide missiles: two hitting the World Trade Centre in New York, leading to the collapse of the famous twin towers, one hitting the Pentagon in Washington DC and the final plane crashing in the countryside in Pennsylvania. The attacks resulted in the deaths of almost 3,000 people with costs to the global economy estimated at over $40 billion (Helminger, 2002). There has been a great deal written about the attacks and most significantly a National Commission into the attacks took place, providing detailed information about them (National Commission on the Terrorist Attacks upon the United States, 2004; abbreviated to '9/11 Commission' hereafter). This information can be used to assess the causes of security failure. The report highlights many issues relating to Al Qaeda in general, to the terrorists, to recommendations and so on, but the focus here will be on the breach of airport screening.

When Mohamed Atta checked in for his flight to Boston from Portland on the morning of 11 September he was picked up by the Computer Assisted Passenger Pre-screening System (CAPPS), as were nine others of the 19 hijackers. This system is designed to identify passengers who should be subject to special security measures – although some passengers are also selected at random – and in this instance the special measures only amounted to holding his checked bags until he was confirmed on the plane. To board American Airlines Flight 11 at Boston the five terrorists had to pass through security checkpoints operated by Globe Security, and for United 175, in another terminal in Boston, Huntleigh USA were performing the screening functions. The purpose of the screening is to identify and confiscate weapons and other prohibited items not allowed on a commercial flight. The screening functions included passing through a metal detector designed to activate on the metal content of a .22 calibre gun. If the detector was activated further screening would take place with a hand wand to identify the objects triggering the alarm. Luggage carried onto the plane would also be X-rayed. The selection by CAPPS of Atta and three others of his team at Boston only affected their treatment by security in terms of their checked baggage. None of the screeners recalled anything unusual about the screening of the terrorists.

At around the same time five more men were boarding American Airlines 77 in Washington Dulles airport. Three of these were flagged by CAPPS and two others were selected for the same additional security measures because of suspicions aroused by the customer service representative, as they lacked photo identification and had very poor English. Again this only resulted in their baggage being held until they were confirmed on the plane. At Washington Dulles, Argenbright Security was responsible for screening and there was CCTV recording the process. Two of these hijackers, Mihdhar and Moqed, set off the detectors and had to pass a second. This time Mihdhar passed and Moqed failed again, so was screened with the wand and then passed. Of the other three only Nawaf al Hazmi set off the detectors, again twice, and was also cleared after the wand was applied. This hijacker was also swiped by an explosive trace detector, which proved negative. Again the screeners did not recall anything unusual. The 9/11 Commission asked a screening expert to assess the hand-wanding. The conclusion was that the quality was 'marginal at best' and that the screeners should have resolved what had set the alarms off (9/11 Commission, 2004: 3).

The final terrorist team were boarding United 93, where two of the four were selected by CAPPS. All four passed through the security screening of Argenbright Security. As at Boston, there were no CCTV cameras, so no footage was available to analyse if any alarms were triggered. Again none of the security staff recalled anything unusual about the terrorists. As the 9/11 Commission concludes on this part of the attacks,

> By 8.00am on the morning of Tuesday September 11, 2001, they had defeated all the security layers that America's civil aviation security system then had in place to prevent a hijacking.
>
> (9/11 Commission, 2004: 4)

The failure of the screeners is demonstrated by the weaponry that the terrorists were carrying. On American 11 attendants and passengers were stabbed, indicating some form of knives, and either mace, pepper spray or some other irritant was also used in the struggle to gain control of the plane, as well as the threat of a bomb. Mace, knives and a bomb were also reported on United 175. On American 77 the hijackers were reported to have knives and box cutters and on United 93 knives and a bomb. The bombs were believed to be fake and the knives were probably of a size undetectable by the metal detectors.

The FAA security regime at the time was focused on the risk of an explosive being placed on a plane (Wilkinson, 2006). Hijacking was not seen as a problem given that there had been no domestic incident of hijacking in a decade. Nevertheless security was supposed to be 'layered', so that if one area failed, there would be another to fall back on. Yet on 11 September, 19 hijackers were able to get onto four planes at three airports.

The layers were intelligence, passenger pre-screening, checkpoint security and onboard security, and they all failed. Intelligence was exposed in the 9/11 report as inadequate, with information held by the state agencies on plots by Middle East terrorists to gain flight training not even passed to airport security authorities. Passenger pre-screening amounted to directions not to fly those who might be a threat to aviation, yet only 12 individuals were on the Federal Aviation Aministration list, when other agencies had lists of thousands of suspected terrorists. The other part of pre-screening, CAPPS, was orientated around the risk of a non-suicide bomb attack utilising luggage in the hold of the aircraft, and resulted in no additional scrutiny of hand baggage. Checkpoint security, the most important 'layer' of all, had been recognised as inadequate by numerous FAA reports and with regular failures of tests by screeners. Add to this the fact that FAA regulations did not expressly prohibit the carrying of knives under 4 inches long and the decision several years before 9/11 to replace random hand searches of carry-on luggage by explosive trace tests. Underlying many of these weaknesses was lobbying by the carriers seeking to minimise the disruption caused by security and to keep costs to a minimum. Finally the onboard security measures were entirely driven by past experience in which hijacking was the precursor to some form of negotiable demands, the risk of suicide had not been considered.

There are many factors that contributed to the overall failure of security on 11 September, but the failure of the screening is an important element. As Wilkinson (2006: 123) argues:

> ...it is important to remember that inadequate and incompetent aviation security at all the US airports involved also contributed to the suicide hijackers' success. Even though the intelligence warning process failed dismally, the coordinated suicide hijackings would have been prevented if the aviation security at the airport boarding gates had been efficient, thorough and comprehensive.

Let us consider what might have happened if the weapons that the terrorists carried on board had been detected and an effective security screener had considered them too menacing (which would have been theoretically possible) (9/11 Commission, 2004: 476).

- Weapons confiscated but the terrorists allowed to board, making it more difficult for them to take control of the plane.
- Weapons confiscated and the terrorists detained for further investigation making the operation for the remaining conspirators (if there were any) more difficult or not possible.
- Weapons discovered and a panic reaction triggered in some of the already nervous hijackers.

Alas these are all hypothetical. All the airports used by the hijackers to start their journeys were category X, the highest level of security. The best security was not a deterrent to these terrorists and it did not stop them (9/11 Commission, 2004: 451). Therefore an analysis of the failure of the screeners exposes systemic issues that contributed to their failings.

* A focus driven by the FAA on potential explosives such that security had little consideration for other new forms of threat.
* Inadequate rules regarding blades.
* A lowcost low-quality norm for security screeners.
* Poor quality screening endemic.
* Too few layers to beat (for this type of attack).

Thus, although it was the security officers who failed to detect the weapons used, the broader systemic issues were the ultimate cause of the screening failure. First, a security regime driven by the bureaucracy of the FAA focused on bombs in unaccompanied baggage rather than hijackings or other potential risks. Second, there were inadequate rules on the carrying of blades that meant there was every chance that if one of the blades of the hijackers had been detected they would still have been allowed to board. Third, in the highly competitive world of aviation, security was regarded grudgingly as a cost to be cut wherever possible. Many companies hired poor calibre guards who were not properly motivated or trained. Poor quality screening was also endemic in the airports, as demonstrated by FAA investigations and many other reports. Finally the system was designed with too few layers to circumvent, even the targeting of 10 of the hijackers by CAPPS triggered no additional layers to overcome (for their type of attack). All the terrorists had to do was get their weapons past *one* screener.

Prince William's birthday party, 21 June 2003

The British Royal Family is supposed to have some of the best security available. Much of their protection is drawn from specialist units in the Metropolitan Police Service. When additional human resources are required these are drawn from what are regarded as highly paid and motivated general police officers from the area in which a royal event is taking place. It was therefore a shock to the world when on 21 June 2003 at Prince William's twenty-first birthday party at Windsor Castle the so-called 'comedy terrorist', Aaron Barschak, dressed as Osama Bin-Laden in a pink dress, managed to breach all security measures and walk into the area where Prince William was speaking to interrupt him and to make his own speech (BBC News, 2003). If this had been a terrorist, stalker or some other more deviant malefactor this might have been a disaster rather than a publicity stunt. It was all the more disconcerting given the high level of security alert because of fears of terrorist attacks at the time.

The subsequent publicity led to a short report into the incident that illustrated numerous lapses in security (City of London Police, 2003). The report found that Barschak had managed to get into the event by climbing two trees and hoisting himself onto a wall from which he could climb down into the castle grounds. This activated an alarm – the first of six alarm activations as he walked towards the prince's party. He changed into his pink dress and Bin Laden outfit in the bushes and on top of the alarm activation was also recorded on CCTV on five occasions. In the grounds Barschak met a contractor and claimed to be a lost guest. The contractor led him to a side door, protected by a police officer, where he was allowed to enter. From here he was able to get to Prince William's party. The report states, 'There appears to have been no operational police response to these alarm activations or the CCTV recordings' (City of London Police, 2003). Other weaknesses included no police presence in the grounds and no police supervision of contractors, either of which would have been able to respond to Barschak. There were 28 recommendations made regarding the incident, relating, among other things, to pre-planning, the policing of the event and to technical issues.

In this incident the security system failed completely. The measures to prevent intrusion failed, the human element in the control room and the guard at the side door also failed. As the report notes one of the key factors was the lack of police presence patrolling the grounds. This alone (although the other recommendations made by the report are entirely appropriate) would probably have been enough to stop this incident occurring. This event therefore points up a flawed security system and inadequacies in the human element.

Modelling security failure

From the analysis above it is possible to begin to model different types of causes of failure. As we have seen above and in Chapter 1 a security system is comprised of human or social and technical elements. There are a multiplicity of different parts of the system that can fail and although on the face of it a failure might seem either a technical or human failure, usually there is some kind of link.

On the technical side there could be a fundamental flaw in the design of the security system. The attack on the Prime Minister at the House of Commons by Fathers 4 Justice highlights a flawed design, with no screen to protect MPs from the risk of missiles thrown from the public gallery. Other examples of design failure are the placement of the security desk at the Gardner Museum and the large numbers of individuals who had access to security knowledge at the Diamond Centre in Antwerp. Layers are also a design issue, but deserve mention in their own right. The weakness of aviation security on 11 September exposed the fact that there were too few layers for the hijackers to get through to get on board the aircraft. Security products can also fail or can simply not be good enough for the task they are intended to perform. For instance, barriers might not be strong enough or an alarm might not work, contributing to security failure. Another common failure is simply that of not anticipating a particular risk, again an issue in the 11 September attacks. All of these failures, however, also have a human element.

Where the human element has been seen to be the first cause of security failure it is often linked to broader systemic failures. In the case of the Brinks Mat robbery, corruption by one security officer seemed the principal cause, but could better vetting have stopped him securing a job? Should there have been additional security where so much gold was being stored? These are legitimate questions about the broader security system. Incompetence may also be an issue, as was demonstrated in both the Gardner heist and at Prince William's party. However, corruption and incompetence might be facilitated by broader problems, such as the recruitment strategy, the training and motivation of security staff and their management and supervision. At the Gardner Museum these could all be seen at work in contributing to the failure. Staff culture is also important and poor practice in airport searching was common in the USA, again contributing to security failure on 9/11. Finally there might not be enough capacity to deal with potential incidents, another factor in play at the prince's party.

Figure 2.1 illustrates common causes of security failure on a continuum where at one extreme is complete technical failure and at the other complete human failure. Most failures are placed in the middle to illustrate that their causes are usually broader systemic failures, rather than one human or product failure. There are some failures that do edge towards complete human or product failure, but there are very few cases that can be located at the extremes of the continuum.

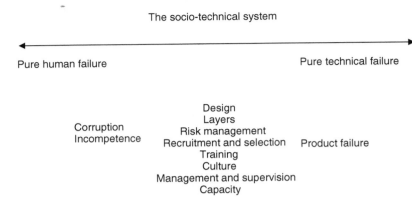

Figure 2.1   Common failures in security systems

Conclusion

This chapter has explored the under-researched subject of security failure and demonstrated that absolute security is neither possible nor often sought and is indeed a myth. The chapter began by examining some examples of security failure as well as illustrating the extent of such failures. It then went on to explore the much more extensively researched area of accidents and disasters and to apply the theories that are devised to explain them to security failures. Three high-profile examples of security failure were examined in order to identify the failures in the socio-technical system that underlay them. Finally the chapter offered a model of security failure, with some of the most common causes of failure.

There are many lessons to be learned here that will be developed in Part III below. Most significantly, many failures do not represent isolated incidents but follow similar or comparable incidents that have occurred previously within an organisation, but which have not been dealt with. Second, there is much to be learned from examples of security failure by other organisations and it is therefore essential that structures should be established to enable informed analysis and debate of security failure to take place outside the glare of publicity. So called 'behind the wire' opportunities should be enabled. This is something that will be developed further in Chapter 9. The security failures examined here also illustrate that the greater the degree of security required the more layers there should be to the system, and that there should be more than one individual making decisions in those layers. The failures of 9/11 illustrate how devastating attacks could be perpetrated by having a system with too few layers that were also known to be ineffective. These ideas will be developed further in Chapter 6. The next chapter explores the views of malefactors on security as this offers priceless insights into how to undo security.

Notes

1 These are estimates and there are inherent problems in the accuracy of crime data, as such they should be treated as a barometer of crime.
2 The House of Commons was debating the abolition of hunting with dogs which had sparked widespread protests all over the UK.

3
Understanding the Malefactors

> If you know your enemy and know yourself you need not fear
> the results of a hundred battles.
>
> (Sun Tzu, *The Art of War*)

Introduction

The defeat of security in order to undertake a spectacular robbery, burglary or fraud has been the subject of many fictional films, dramas and
books. Probably the most famous of such movies of the last few years is
the 'Ocean's Eleven' (2001) series ('Ocean's Twelve' (2004) and
'Ocean's Thirteen' (2007) quickly followed the success of the first film)
in which Danny Ocean puts together a team of villains, all with expert
skills, to undertake spectacular heists. These are recent examples of a
genre with a long history. Equally dramatic heists are the theme of
'The Italian Job' (1969; with a re-make in 2003), 'Heist' (2001) and 'The
Ladykillers' (1955; re-made by the Coen brothers in 2004) to name
only a few. Despite the substantial exploitation of such ventures
in fiction, there has been relatively little academic research either on
those who undertake such major crimes or on the perpetrators of
more mundane criminal acts in real life. This chapter aims to gain
a greater understanding of malefactors in some of the most common areas of criminal behaviour. It will attempt to assess their
decision-making strategies and in particular their views on security. It
will consider the common crimes of theft, burglary, fraud and robbery
as well as protest. As the famous quotation from Sun Tzu illustrates,
knowing the enemy (malefactors in this case) brings benefits in defeating them.

Academic research on malefactors has been neglected. As Gill (2005a: 306) has argued,

Newcomers to criminology, and specifically the study of crime prevention/reduction and criminal justice policy, may be surprised to learn that the subject has not been dominated by studying offenders and gaining insights from them to inform policy.

Researchers such as Ekblom (2005: 203) have urged designers to 'think thief', and Gill (2005a) has championed the need to understand the decision-making strategies of criminals in order to develop effective security. Consequently there is a growing base of literature to draw upon (Bennett and Wright, 1984; Butler, 1994; Gill and Mathews, 1994; Gill, 2000, 2005b). There have also been numerous more recent accounts in the literature and other media by ex-malefactors of how and why they went about pursuing their crime or 'campaign' (Leeson, 1996; O'Conner, 2007). Finally there are numerous secondary accounts in newspapers, books, television and so on speculating – on the basis of varying amounts of evidence – as to how criminal and other incidents were planned. Although it is important to note the caveats of this type of research, and particularly the accounts of offenders, who may lie, embellish their stories and minimise their guilt (Gill, 2005a). Bearing these in mind, however, there is still much to learn from them. Under the main heading of 'Opportunistic malefactors' the chapter will therefore begin with general explanations of offending behaviour before moving on to specific crimes and types of incidents. It will then examine the different types of malefactors, the places that they are likely to target and the risks that they pose. A case study of the television programme, 'The Heist', which set former professional criminals the challenge of breaching different security systems is used to illustrate the ways in which 'determined malefactors' go about their work.

Opportunistic malefactors

There has been a great deal of research to illustrate the extent to which crime is related to opportunity (see for example, Mayhew et al., 1976; Clarke, 1980, 1992). During the 1970s in the UK, a whole series of Home Office research studies, inspired by data on suicides, pointed towards the importance of opportunity in crime. When toxic town gas was replaced by natural gas, the preferred method of suicide (of switching on the gas with one's head in the oven) disappeared. However, research discovered that rather than using another method many chose not to take their lives and the total number of suicides actually declined. A similar effect was also found in the USA with the introduction of catalytic converters on cars. It was therefore argued that if as fundamental a decision as to take one's life was based upon opportunity, then the much lesser decision of committing a crime may also be similarly affected (Clarke and Mayhew, 1988). This led to extensive research into opportunity reduction, which showed amongst many other things that the introduction of the compulsory wearing of helmets for motorcyclists led to a reduction in the number of motorbike thefts and that compulsory steering locks in cars in Germany reduced the number of car thefts (see Mayhew et al., 1976). Alongside these ideas of rational choice theory, routine activity theory emerged. This argued that three elements are required for a crime to occur: a motivated offender, a suitable victim and the absence of a capable guardian (Cohen and Felson, 1979). Gill (2005a) has also highlighted the fact that the motivated offender requires the appropriate resources – and there also needs to be a law to break.

The above is only a very brief précis of some of the extensive research in this area. Nevertheless, acceptance of the significance of opportunity in levels of crime has led to further research on the benefits of situational crime prevention techniques, something that will be explored in more depth in Chapter 7. The focus here, however, is on the malefactor. Many malefactors will commit a crime because at a suitable target (victim) there is an opportunity to do so (absence of capable guardian). These opportunistic offenders or malefactors are denizens who are not normally offenders, but commit a crime because of some combination of personal circumstances and opportunity. There are other malefactors who may actively seek out opportunities or who are simply not concerned with being caught. The following sections will review a range of criminal acts and other deviance (as defined by various nodes) and examine the research and literature on malefactors' decision-making strategies, with particular reference to their views on security.

Shop-theft

In 2005–06, the British Retail Consortium estimated that retail crime cost retailers over £2.1 billion (including costs of security) of which a significant amount related to shop-theft (British Retail Consortium, 2007). There have, however, been no more than a handful of studies that have sought more information on shoplifters' decision-making strategies. Butler (1994) used 15 shoplifters in role plays where they were given the task of stealing a video recorder from a choice of three stores in a high street, a superstore or a shopping mall. None chose the shopping mall, with nine choosing the superstore and six the high street. The mall location was rejected on the grounds of limited access and therefore lack of escape routes. In terms of security measures the biggest risk to the shoplifters was being apprehended by a member of staff, second was being tackled by a 'have a go hero'. Ten stated that they would not be deterred by security guards generally, but would be deterred if the guards followed them. Similarly if ordinary staff were nearby, nine would be deterred. Eight were deterred by security cabinets or chains on the goods and seven by CCTV, or goods linked to an alarm or if a store had a photograph of them relating to previous offending. As well as the role plays Butler conducted interviews with the shoplifters and these reinforced the perception that many security strategies, such as loop alarms and chains, had marginal deterrent impact. As one interviewee told Butler (1994: 68) about the deterrent effect of alarms,

'No, it wouldn't put me off. If I was going to do it properly…if I was going to take something off that shelf there, you come in with a small pair of pliers, lift the unit up and cut the wires. Sod the alarm.'

And if the alarm went off? He responded,

'Grab it and run out of the fucking door.'

The most significant factors for the shoplifters were the ability to escape and the desire to avoid being apprehended. In security terms this meant staff being nearby or security observing their actions.

In more recent research on shoplifters offering an international perspective from Brazil, Canada, Spain, the USA and the UK further insights in the minds of shoplifters were found (Gill, 2007). Six stages in the decision-making process were identified: choosing the store, entering the store, locating the product, taking the product, leaving the store and disposing of the goods.

Gill (2007) found that most shoplifters preferred to steal locally and not to travel too far. They also liked familiarity, such as stores where they had successfully stolen before or national chains with familiar organisation. The thieves also liked disorganisation in the store, which would enable them to blend into the background. Gill found security was a factor, with those stores with a reputation for good security being shunned. He also found that thieves made some investigation into the level of security in a store, preferring to steal from shops with low levels of security. The shoplifters were also attracted to stores where they knew the staff had low levels of commitment. Some of the offenders also preferred stores that were busy, which gave them cover for their crimes. As in the earlier research, escape routes were a very important factor. Some chains of stores were seen as more generally exploitative of staff and customers which helped to rationalise targeting. The final factor related to the types of goods required.

Gill (2007) found that on entering the store offenders would first assess whether they were likely to be seen; it was felt important not to stand out and to fit in with ordinary customers. They would also assess the level of security and the location of cameras and try to identify any blind spots. Again a very important factor at this stage was to work out any escape routes. Some shoplifters steal to order and would head straight for the products sought, others would search for items they knew they could easily sell, such as fast moving consumer goods (batteries, razors and so on).

When it came to concealing the product some shoplifters would place the booty under their ordinary clothing; others would wear special clothes, such as coats with large internal pockets. Some offenders would hide goods closer to their bodies, for example concealing goods under their armpits or in more intimate places. Some offenders used specially designed packages to conceal goods, such as an empty package wrapped to resemble a gift. An umbrella could also be used to drop small items in. Special bags can also be used, some of which are designed to stop tags activating electronic article surveillance (EAS) alarms. Some shoplifters simply use ordinary bags or shopping trolleys to conceal goods. Another strategy was 'steaming', where a large group of people act together to simply run off with the goods. Some offenders have magician-like abilities enabling them quickly to hide goods, and others may use distraction. Collusion with insiders was another common strategy. The exit was also a very important factor in the decision-making strategy. Some would blend in by purchasing some goods, while concealing others. The thieves were very aware of their surroundings and if they felt they were being observed or followed they would abandon the goods.

Generally Gill's (2007) study found that rather than deterring shoplifters, security measures by the stores seemed to be perceived as presenting additional obstacles that needed to be overcome. EAS tags were not rated very highly by thieves in the study because frequently the activations were ignored by staff. Other strategies included destroying the tag, removing the tag, placing the goods in a magic bag, deactivating the alarms, or circumventing the gate. CCTV was viewed as a major threat by shoplifters, but again there were strategies that could be used to address this risk. Some of these included the skill in the act of theft itself, such as sleight of hand. Others included checking the line of sight of the cameras and finding blind-spots or the exploitation of crowds to conceal activities. Being watched or followed by security guards was seen as a particular threat in Spain (which has some of the highest regulatory standards in the world) but elsewhere it was reasoned that many guards were poorly trained and motivated and were not likely to pose a threat. Nonetheless, there were also strategies to deal with security guards, such as simply aborting the theft, relying on the guard's incompetence, bribing the guard, and threatening guards in areas where they are thin on the ground. Such areas might also be targeted by thieves. The importance of surveillance was central to influencing shoplifters' decisions, Gill (2007: 29) argued, '...shop thieves do not like being watched; surveillance is a powerful deterrent'. The conclusion Gill (2007: 33; italics in the original) came to on the effectiveness of security was,

> ...mirroring studies of security generally, is that all measures have a value; they all have potential to work in some cases against some kinds of offenders and offences. And the absence of measures was definitely an advantage:
>
> *OK, wow... wow... wow...this is probably the easiest one to steal out of that I've been here [to]. No cameras, no mirrors, no people, no staff.*

Gill also identified three key areas that the shoplifters rely on for success. These were inappropriate and poorly managed security measures, the failure to support staff with training or procedures and insufficient attention to design issues. These are clearly areas the security decision-maker can work on to make security a more potent threat to malefactors.

Burglary

There have been a number of studies on burglary, most of them relating to domestic burglars. These studies offer some insights into decision-making strategies and in particular the views of burglars on security as a barrier to committing their crimes. In one of the first studies of burglars Maguire and Bennett (1982) identified three types of burglar:

- High Level: a small network of highly skilled criminals, who are well organised and socially exclusive.

- Middle Range: the most ubiquitous burglar who is skilled but not so well socially organised.
- Low Level: petty criminals pursuing indiscriminate crimes with limited skills and organisation.

In their study of decision-making strategies, Maguire and Bennett found that security measures based upon target hardening were not that effective. Indeed, in their profile of Peter Hudson, whose criminal career had culminated in burgling country houses, Maguire and Bennett (1982: 102) found that 'Security measures rarely presented any problems to him. Indeed, he took a perverse pride in beating them.' Further research by Bennett and Wright (1984: 95) based on interviews and experiments with 300 convicted burglars found that,

> Although security locks and levels of security were mentioned by burglars, they were not mentioned very frequently and they were not described as being particularly influential determinants in their decision to offend.

Maguire and Bennett (1982) also found that much of the information that enabled burglaries to take place came from people overheard or talking unguardedly about valuables or security procedures. In a more recent study of 50 burglars, Nee and Meenaghan (2006) found that 39 were motivated by the need for money, with six motivated by the excitement of the crime and five claiming the influence of others as the main motivation. Three-quarters of the sample made their decision to commit the burglary away from the scene of the crime and then travelled to a suitable area to find an appropriate target. In their selection techniques the most important factors were signs of wealth and layout. In terms of the latter, the degree of cover, access and potential 'get away' routes were all very important. Security, however, was not at the top of burglars' lists in selection strategies. Eleven stated they always avoided dogs and another 11 sometimes avoided dogs, but 30 claimed that they were not deterred by either dogs or alarms and 21 felt security had improved over the last 10 years. However, Nee and Meenaghan (2006: 942) noted:

> In any case, all participants felt security features were rarely enough to deter them, due to lack of vigilance in locking up on the part of

homeowners. The most common reason for abandoning a recent burglary was because they had been disturbed (22) rather than being deterred by insurmountable security (14).

Other studies, however, have come to different conclusions on the impact of security measures. For example Hakim and Buck (1991) found that an intruder alarm was a good deterrent against burglary. There is also the impact an alarm has if activated during a burglary. Conklin and Bittner (1973) found that in 75 per cent of burglaries where an alarm was activated no goods were stolen. More recent research by Lee (2006) on South Korean burglars found that 82.7 per cent of burglars studied were deterred by an intruder alarm. He also found 84.6 per cent deterred by CCTV and 75 per cent by dogs. Less of a deterrent were locks, high windows and high fences/walls. Lee also found that the deterrent effect of all of these measures were greater on younger than older burglars. Lee concluded that the effectiveness of security strategies in deterring burglars may rise the more of them there are. He wrote, '...the more obstacles the burglar would have to overcome, the greater the chances that they might decide not to undertake the planned burglary attempt' (Lee, 2006: 142).

Fraud

Fraudsters have been estimated to cost the UK economy around £14 billion per annum (Levi et al., 2007). In the USA occupational fraud alone has been estimated at costing an annual $660 billion and 30 per cent of small business failures have been attributed to employee theft (Association of Certified Fraud Examiners, 2004; Kuratko et al., 2000). Fraud encompasses a wide range of different types of deviant behaviour, from social security claimants making false claims about their personal circumstances to scams involving 'fake' corrupt Nigerian ministers to staff defrauding their own companies. Of the latter type of fraud there have been a few studies that have sought to profile the fraudsters and understand their motivations and modus operandi.

KPMG Forensic's (2007) survey is based on 360 cases of fraud from 700 investigations in Europe, the Middle East and Africa. Their typical fraudster was a male (85 per cent), middle aged (between 36 and 55) (70 per cent), committing a fraud against his own employer (89 per cent), and acting alone (68 per cent). Over 50 per cent of fraudsters had worked for their employer for at least six years. Most disturbingly two-thirds of all primary perpetrators were from top management. The research found that greed and opportunism accounted for 73 per cent of fraudsters and that in more than half the cases there had been no prior suspicion. In 49 per cent of cases the fraudsters had been able to exploit weak internal controls. The report argues that there are three main factors connected to the committing of fraud. First is opportunity provided by weak internal controls (security measures). Second is motive, where the way of life of the perpetrator leads to financial pressures that culminate in the commission of fraud to fill the gap between salary and lifestyle. Finally there is rationalisation, where the fraudster justifies the fraud on the basis that they are in some sense 'owed' money by their employer.

In another study, based on interviews with 16 convicted fraudsters of 'large sums' of money ranging from £65,000 to £25 million, it was found that debts, greed, boredom, blackmail, temporary insanity, desire for status and a corrupt company culture were some of the motivations for fraud (Gill, 2005b). Central to the fraudsters' accounts of how they were able to commit their frauds were weak controls (poor security). This amounted to the exploitation of systems of checking that only reviewed a small sample of transactions under a specified amount, £750 for example, making it easy to steal sums up to this level. Limited capacity for overseeing transactions was another problem; in one case a single accountant and temporary staff oversaw invoices and these were never checked. In another case a fraudster oversaw a £20 million budget and could – alone – sign invoices up to £25,000; he created companies invoicing sums below this amount for non-existent services. The survey also revealed the weakness of auditing arrangements in discovering crimes. It is worth quoting the views of one fraudster, Eric, who defrauded his company of several million pounds, on the auditors.

> Accountants can only work on the figures they have got, audit the same. Auditors came to see me and I just lied to them and gave them false pieces of paper and that was that. The checking process was abysmal. I was not worried because I have 20 years experience of auditors. Had they been better at their job I would have been in trouble. What I was doing was simple, but the lack of process enabled me to do what I did, the absence of systems, the lack of attention to detail, the lack of knowledge in auditing and accounting. I had three audits in those 18 months ... I gave the auditor the information and he said, 'thank goodness for that', and my thought was, 'you complete

muppet'...There was no interrogation from audit and that was good for me.

(Gill, 2005b: 40)

It is interesting to note that most of the offenders did not consider what they were doing to be high risk, nor did they consider that they were likely to get caught. However, all 16 interviewees had been caught, and for a variety of reasons. Some were caught as a result of external investigations, others because new members of staff moved into positions of control, new auditors were appointed or anomalies eventually emerged in accounts.

Kapardis and Krambia-Kapardis (2004) draw upon numerous empirical studies of fraud to identify 12 types of offenders. Some of these can be grouped together to create eight categories. These are listed below with the percentage of cases in brackets:

- Predator/career fraudster (16 per cent) or serial fraud by unscrupulous deceiver (14 per cent)
- Opportunistic first offender in professional occupation (24 per cent)
- Fraud under an assumed professional identity (2 per cent)
- Isolated fraud as response to unsharable financial pressure on family (4 per cent) or oneself (4 per cent)
- Serial fraud as response to unsharable financial pressure on family (10 per cent), oneself (8 per cent) or loved ones (8 per cent)
- Fraud as personal justice (2 per cent)
- Serial fraud due to vice (6 per cent)
- Isolated fraud to restore social identity (2 per cent)

Kapardis and Krambia-Kapardis (2004: 197) were able to identify that the majority of serious fraud offences were perpetrated by males, 35 to 45 years old, of high educational status, either greedy or with a serious financial problem, with no prior criminal record, in positions of trust, who rationalise their behaviour, specialise in fraud, act alone, use false documents, victimise two or more people they know and are convicted of multiple charges. They also concluded that frauds are rarely isolated events as many will not stop until they are found out. Therefore in cases where controls are weak and the fraudster is unlikely to be found out, strong accounting controls and security procedures could have an impact upon the incidence of fraud. They argued (2004: 197) that,

> ... effective preventative measures against fraud are undoubtedly strong internal accounting controls and stringent staff employment-screening procedures to minimise opportunities for fraud to occur, and better vetting of job applicants by employers to screen out the predator [sic], career fraud offender with a criminal record, often for deception offences...

However, the majority did not have a criminal record before their fraud was discovered, and what was more important was the opportunity to commit fraud, the crises in the lives of the fraudsters and the organisational culture within which they worked. Indeed it is interesting to note that the collapse of Barings Bank caused by the £850 million fraud by Nick Leeson started with a £20,000 fraud to cover a fellow employee's mistake. The controls, management and auditing arrangements of Barings were extremely poor and at one stage, when the losses were relatively manageable by Barings, Leeson could have been stopped in his tracks (Leeson, 1996: 93). Instead over three years he was able to build up losses that culminated in the collapse of the bank. At a conference I attended on private investigators not long after the collapse a private investigator advocating more staff-vetting showed how, on hearing of the collapse of the bank and the name of the suspected culprit, he was able to secure financial information about Leeson which would have led to his never having been offered a position. Such vetting (which was not undertaken) cost a few hundred pounds and took a very few minutes!

Another rationale for employee fraud which has become more prominent in recent years is workplace dissatisfaction and workplace equality complaints (Hollinger and Davis, 2006). This research suggests that the more positive employees' views are towards their organisation, management, supervisors and so on the less workplace deviance there is. Therefore measures to combat issues that promote negative attitudes such as inadequate or unequal pay, poor working conditions, little staff participation, poor workplace relations and the like can also have a positive impact.

Robbery

There have been no academic studies of robbers engaged in large well-organised robberies that net hundreds of thousands or even millions of pounds worth of loot. There have, however, been studies of smaller-scale commercial robberies. Gill and Mathews (1994) undertook a study of 341 convicted robbers of commercial premises. All but one were male, over half were in their 20s and over 80 per cent had received their first conviction before they were 21. Gill and Mathews found that the decision to commit a robbery was most frequently based on unemployment, and the desire for money, excitement and drugs. Central to most was a lifestyle involving drugs and alcohol and the need for cash to fund this habit. Preparation for robberies was generally not that extensive with variations between different types of target. For example, for building societies 60 per cent of robbers spent up to a day preparing their crime, whereas for cash-in-transit robberies 12.8 per cent spent up to a day in preparation and 63.8 per cent took the longest preparation time, of more than a week. The length of time that some robbers are prepared to take is illustrated by the Great Train Robbery of 1963 which netted £2.5 million and was planned in depth over several months if not years (Fordham, 1968). Later research by Gill (2000) also distinguished between the 'professional' and 'amateur' robber depending upon their degree of planning, their use of the loot and the excitement they derived from it. Clearly the different orientation also has implications for the potential impact of security. 'Professionals' who invest more time in planning would be much more likely to target riskier targets, seeking to use their expertise to overcome any security barriers. 'Amateurs' would be less likely to target a riskier location, preferring to select one that is 'easier'.

The survey also sought information on the robbers' assessment of security measures. Many offered limited insight on their decision other than 'instinct' or 'it looked good'. However the authors did find that ease of escape was a significant factor in target selection. They also found the presence of CCTV offered little deterrent as disguises would be worn. Alarms also offered little deterrent. Some later research confirmed that, 'For the most part though the presence of cameras was taken for granted [by robbers]' (Gill, 2000: 49). Rising shutters proved a more significant factor in decision-making as they render the escape with loot more difficult, although some robbers saw an advantage in the inability of staff to 'have a go' at them. Gill and Mathews (1994: 26) offered the following conclusion from their research, 'The greater the risk – from the robber's perspective – the less likely it is he/she will proceed. Security measures have a part to play here, and so do other things which may frustrate an escape.' Nevertheless as Gill (2000: 58) argues in his later work it would be wrong to think security provides the answer to effectively preventing all robberies as '...some robbers stole for very little money and were oblivious to the risks involved. And some planned their offence carefully or were so determined that it would take a fortress to thwart them'.

Another important issue in robbery is the need for information. Robbers relied on a number of strategies to secure appropriate 'knowledge'. Many staff are indiscreet and robbers can secure information simply by the accidental leaking of sensitive knowledge. In other situations robbers may simply attempt to bribe staff and security to give appropriate information or use more deviant tactics, such as blackmail or sheer force. The other strategy often used is for someone with criminal intent to secure a job on the 'inside'. Much of this leaking of information is preventable by having appropriate policies and training in place on the release and discussion of security data and, even more importantly, on the recruitment and vetting of staff. This is something that will be returned to in depth in Chapter 8.

Protest

Protesters encompass a very wide range of people from a variety of social backgrounds. The vast majority of protesters act within the law and pose no threat to security. There are, however, a number of issues that have spawned more radical protesters who have been prepared to engage in direct action and break the law. The issues of the environment, animal rights, hunting and fathers' access to children are some of the most prominent examples. The skills of some environmental protesters led me to describe it as the professionalisation of protest (Button et al., 2002). Given the problems of access and the nature of many protesters' activities, academic studies using the same methodology as some of the offender studies cited above have not been undertaken on protesters. However, there are numerous guides for protesters and some published accounts of their exploits. Some of these will briefly be explored to examine their views on security.

In, *Road Raging Top Tips for Wrecking Roadbuilding* there is a whole section on 'knowing the enemy', which includes politicians, road-building agencies, financers, security guards, private investigators and the police (Road Alert, 1997). Of particular interest are the insights on security guards, who are seen as a group open to 'defecting' to their cause, as potential sources of intelligence and with low commitment to their employer.

There are many examples that illustrate how protesters have undermined security to further their causes. For example during the protests against the Newbury bypass (where over a thousand security officers were employed at some periods) protesters noted that 300 of the guards were billeted on a farm accessed via a narrow lane. One day the protesters got to the lane before dawn and raised scaffolding pole tripods which they sat on at either end of the lane, while the guards were trapped in their coaches in between. That day, no security got to the protest site (Road Alert, 1997). Indeed, it is suggested that a general over-reliance on vehicles can be used by protesters to scupper security plans. As the guide states, 'Always seek to ambush and wrong-foot contractors before they get to you. Their plans are complex and their organisations rigid, so a little spanner in the works here and there can spoil their plans beautifully!' (Road Alert, 1997: 85).

In environmental protests site invasions and occupations are a popular strategy. Again the guidebook suggests using decoys, approaching sites from different directions and invading/occupying multiple sites in order to split what is often a 'thin' security presence. The police and security frequently employ evidence-gathering teams who use camcorders and cameras. The guide offers detailed advice on how best to disrupt this through the use of banners and the location of protesters.

Another frequently-used tactic is the occupation of office buildings. This requires breaching the access controls of an organisation. The guide states,

> One smartly-dressed person going in and opening the door for everyone else often works. Small bits of wood can wedge the door open. Have an excuse such as courier delivery, an employment enquiry or a pre-booked appointment with a named worker. Alternatively, you can sneak in behind employees going in, or catch the door as one leaves.
>
> (Road Alert, 1997: 93)

There have been very few accounts of how protesters have undertaken their daring deeds. One of the few is Matt O'Conner's (2007) account of Fathers 4 Justice. Unfortunately he does not give a great deal of detail on their analysis of the security systems they had to beat. He does set out, however, that,

> Every Fathers 4 Justice protest was meticulously planned. They still went tits up, but they were meticulously planned. Whilst many newspapers had F4J down as a slickly run military machine, in

reality it was shambollocks [sic]. Everything seemed to happen in spite of our precision planning, not because of it, and our first piece of high wire work was no exception.

(O'Conner, 2007: 77)

O'Conner reveals how lucky the protesters were on a number of occasions. For example, in one stunt in London where a protester climbed up a crane, with the result that roads were closed for five days at an estimated £9 million cost, the protesters were able to get to the crane through a pre-existing hole in the fence! In probably their most famous stunt – where protesters threw condoms filled with purple flour at the Prime Minister in the House of Commons – he reveals that while their many visits to the Palace of Westminster had shown how tight security was, it had also shown them that no intimate searches took place so that all they had to do to breach security was to place the condom in an intimate place. The missiles could be easily removed and thrown from the public galleries of the chamber, where the barrier was not yet complete and no further searches of 'strangers' take place. In the other infamous stunt where a protester dressed as Batman was able to get onto the ledge of the balcony of Buckingham Palace, numerous reconnaissance missions had shown them the gaps in security around the Palace. Underlying their success in beating security was research (of varying degrees of sophistication) into the target, planning, practice and luck.

Malefactors

Earlier in this chapter we examined the importance of opportunity in criminal events. Many malefactors are ordinary denizens suddenly confronted with an opportunity to commit a crime; they have not set out to commit a crime, but are prepared to take advantage of the opportunity for personal reasons. Opportunist malefactors are the easiest challenge for security, for if opportunity can be minimised, so will the deviant behaviour that it provokes. From the discussion above it seemed that fraudsters in particular fitted this category, but so did some shoplifters, burglars and robbers. However, there are many other types of malefactor who might make active attempts to identify an opportunity or to whom opportunity is not a concern. These can be termed the 'D' malefactors, because their descriptions all begin with the letter D. These malefactors pose the greatest challenge for security.

Determined malefactors

Determined malefactors include a wide range of groups. They are organised and committed to breaching the security of a particular node. They include organised criminal gangs, terrorist groups and protesters.

Organised criminals

Organised criminals work in large-scale highly organised groups – the mafia, triads and so on – and engage in a wide range of criminal activities. For the purpose of this book the term will also be used to cover other criminals, whether lone or in groups, who possess a high degree of expertise and look to regular criminal activities to secure an income. Among the more common crimes undertaken by this category of malefactors are robbery, burglary, fraud, drug crime and running prostitution, but material gain is the central focus. The Great Train Robbers exemplify the type of determined malefactors, taking months to plan and organise their infamous heist (Fordham, 1968). Equally determined are the 'professional' cat burglar Peter Hudson, whose story is told by Maguire and Bennett (1982), and the 'high level' and 'middle range' burglars they identified. For this type of malefactor security is viewed as a hurdle to be overcome through planning, bribery and in some cases physical violence. Underlying all these strategies, however, will be the desire not to get caught, which ultimately places some limits on what they are capable of.

Terrorists

In terms of definition, terrorists are the subject of much debate, but for the purposes of this book they will be considered as individuals or groups who pursue campaigns of violence motivated by political demands. Their risk profile focuses more upon acts of violence, although materialistic crime may also be pursued to fund activities. 'Traditional' terrorist groups such as the PIRA can be distinguished from the so called 'new terrorism' (Wilkinson, 2006). Networks such as Al Qaeda differ significantly in their aims and strategies from 'traditional' terrorists. Some of the characteristics that distinguish 'new' from 'traditional' terrorism include:

- a determination to inflict mass casualties on innocent civilians
- the willingness of perpetrators to kill themselves as well as their victims during an attack
- an increased threat of the use of weapons of mass destruction, such as chemical, biological, nuclear or radiological (CBNR) weapons, and

- the pursuit of radical, opaque, non-negotiable objectives (House of Commons Defence Committee, 2001).

As with organised criminals traditional terrorists will view security as an obstacle and will seek to plan their way round it, bribe individuals or use violence. With new terrorism the threat is higher because not only will some be prepared to be caught, but some will also be prepared to die. Thus their rationale for dealing with security includes the possibility of direct confrontation, regardless of risk.

Protesters

These are groups campaigning on a political issue who are prepared to pursue direct action to further their campaign. Their risk profile may include stunts, occupations, and acts of intimidation and sabotage. Like organised criminals and terrorists they view security as a hurdle and will plan to circumvent it. They are unlikely to use bribery, although intimidation and physical force might be used, with more serious violence very rare. The orientation, however, does depend upon the type of group. For example radical pressure groups such as Greenpeace and Friends of the Earth would be prepared to engage in direct action, but not violence, whereas more radical militant environmental activists would be prepared to pursue criminal damage, intimidation and so on (Button et al., 2002). O'Conner (2007: 78) in his account of Fathers 4 Justice provides a more anecdotal analysis of types of protester. At one end of the spectrum is the 'model citizen' who is willing to participate in demonstrations, but not direct action. Next is the 'armchair general', keen on the idea of protest but reluctant to get involved. The third is the 'reluctant enthusiast' who is prepared to be involved in planning, but not prepared to be involved on the front line. Fourth is the 'borderline activist' who is ready to get involved if necessary in a support capacity. Finally he identified the 'fully certifiable activist' who is a hundred per cent committed.

Drunk and drugged malefactors

When people are drunk or high on drugs they lose their inhibitions and the ability to make rational decisions. For this group crimes might be attempted even if there is no obvious opportunity. Many crimes of violence are perpetrated by malefactors who are drunk or under the influence of drugs (Licu and Fisher, 2006). Some of the robbers described by Gill above were so driven by their need for drink and/or drugs that they had no fear of either security systems or the potential consequences of their acts when conducting their robberies.

Deranged and dumb malefactors

Similarly there are people who have mental health problems or who are simply – to use the colloquial expression – so dumb that circumstances might lead them to pursue deviant behaviour whether there is an opportunity or not. The murder of Dr Erlina Ursua (discussed in Chapter 2) by a mentally ill patient demonstrates this.

Desperate malefactors

Finally there are the desperate, who are prepared to take risks in an attempt to ameliorate their situation. An example might be a drug addict desperate for cash or a person determined to escape a country. At the height of the Cold War there were many people so desperate to escape Communism that they took major risks to breach the security systems holding them in the East, often tragically. In August 1962 Peter Fechter tried to escape to the West; shot by East German border guards he was left to bleed to death in front of the world's media.

Now that we have considered some of the tactics and types of malefactors, it is worth examining a case study of determined malefactors in action to illustrate how they go about their work.

Case study: 'The Heist', simulating determined malefactors

In December 2004 the British television Channel 4 broadcast three episodes of 'The Heist'. This programme used former professional criminals to simulate the undertaking of a crime. The reformed criminals included Joey, previously an organised criminal; Jerry, a convicted armed robber; Mathew, a computer hacker; Arnold, a German extortionist; and Peter, a former 'cat burglar'. In 'Art Attack' they were set the challenge of stealing a painting from an art fair at the London Business Design Centre; in 'Million Pound Car' they were tasked with stealing an experimental prototype car; and finally in 'Horse Trading' they had to kidnap a horse and secure a ransom for it. These were programmes designed for entertainment, but where they are useful is in demonstrating how professional criminals go about planning a complex heist, the factors they consider and the weaknesses they exploit.

In 'Art Attack' the security manager in charge of protecting the targeted artwork was happy for the security to be tested, seeing it as a means of pointing up weaknesses and thus instituting improvements.

The only rule for the team was that no violence was allowed, and for the security team there were to be no additional security measures beyond those already established. The security manager was very confident, describing the centre as 'Fort Knox'. There was a 24-hour security presence, extensive CCTV, access control procedures and so on.

The team had four days, and their first move was to attend the fair and explore opportunities to steal the painting. They were given hidden cameras by the programme makers to show the strategies that they pursued. Meetings were held with all the team, Joey chairing, where weaknesses and potential strategies were debated and actions agreed. The team explored the centre extensively, including the loading bays and escape routes, which were central to their analysis. They observed that the painting was located near an exit, and Peter discovered that some windows were only secured by bolts and that there was no CCTV coverage outside them. The team decided on the following strategy. Jerry was to go in the night before as a normal visitor but he would smuggle in a small cutter to cut the bolts on the window, leaving them with the appearance of being shut. However, he was thwarted in this by a security guard who was sitting on the stairs nearby reading a newspaper. It was impossible to attempt to cut the bolts because of the potential for noise. The guard stayed there for several hours, so Jerry had to abort this plan and just leave the windows open, hoping that security was lax and they would not be closed.

The next stage of the heist was for Jerry and Arnold to turn up at the windows (which would hopefully still be unbolted), posing as window cleaners. Arnold was then to climb into the centre, grab the painting and make his escape using the ladders and the waiting 'window cleaners' van'. Unfortunately the bolts were locked, but Jerry was able to force the windows with little effort. Arnold then sneaked into the area where the painting was, removed it and walked out. The team had successfully stolen the painting. The second part of their task, negotiating a reward for its return, was less successful.

In the 'Million Pound Car' the team had to steal a prototype TVR and smuggle it to Belgium. The car was to be at a motor-show in Earls Court, London and was stored in an old aircraft hangar outside London. There was extensive security at Earls Court and at the hangar, and there was a tracking device on the car. Again the team made careful surveys of both Earls Court and the airfield where the hangar was. At the airfield they discovered that security was indeed tight at the front, but at the back there was a gate with no protection. They also realised that the weakest point in the security was the transportation phase. They decided to set up a fake security post at the back gate and to use a ruse to get the driver of the lorry out of his cab. The hacker was tasked with sending a memo to the driver informing him of a new route to the hangar which would take him to the back entrance to the site. The team purchased all the equipment to simulate a security post from high-street shops and set it up minutes before the lorry arrived. They also placed their own tracking device on the lorry, so they knew where it would be. When the lorry arrived at the gate the 'security guard' (Joey's nephew) approached the driver with a clipboard; at that point a fake phone bell rang and he went back to the security hut, returning and telling the lorry driver 'it's for you'. The driver got out of the cab and went into the hut. In a real heist at this point he would have been overcome and tied up, but as this was television a member of the production team was there to tell him what was going on and ensure that he played along. The hacker then located and removed the tracking device from the car and it was driven off. It was also successfully transferred to Belgium. In this case the team were completely successful.

In the final episode, 'Horse Trading' the main challenge was to secure a ransom for a kidnapped race horse. In this case the horse was stolen easily as a stable boy on the 'inside' simply led him out to the gang. The main challenge here was to get the money from the owner without getting caught, which is not as relevant to this discussion. They were successful in securing the money, but it was rendered worthless when a security device exploded dye over it.

Again we should reiterate that these were simulations intended as entertainment. What they are useful in demonstrating, however, are the decision-making strategies of these 'determined malefactors'. First, they undertook extensive planning. In each case they identified the weak links in the broader security systems. In the 'Art Theft' it was an ineffective window where there was no surveillance and in the 'Million Pound Car' it was the transportation element and the unguarded entrance to the airfield. The fewest and the weakest security layers were the target of the gang. Security measures were avoided if possible, or viewed as obstacles to be overcome (such as the tracker on the car). In 'Art Attack', the security manager at the Business Design Centre professed to be pleased with the outcome as it enabled him to close gaps in security of which he had not been aware. Indeed such simulations should be used more frequently by security managers; something I shall return to in Chapter 8.

Conclusion

Research on malefactors and their attitudes to security is a subject in need of further attention. This chapter has built upon some of the small number of academic studies and other relevant literature to illustrate how malefactors go about their 'work'. In the first part it explored their tactics and what they thought about security. It showed that, in most cases, for shop-theft, burglary, fraud, robbery and protest, security was not regarded as a major threat to the commission of a crime or its weakness had been a factor in the success of a crime. Surveillance was influential in shop-theft, but, surprisingly, not for robbery. Far more important for theft, burglary and robbery was the ability to escape. In the second part the chapter assessed the different type of malefactors. It demonstrated that there are genuinely opportunistic malefactors, as well as D malefactors, the determined (organised criminals, terrorists and protesters), the drugged and drunk, the deranged and dumb and the desperate. For the D malefactors opportunity could also be of significance although it might be actively sought, but in other situations the state of mind of the malefactor was such that opportunity was not a factor. This again has implications for security; for the D malefactors it needs to be good enough not only to deter, where that is possible, but also to intervene where necessary.

Finally the Channel 4 series, 'The Heist', which simulated the planning and execution of spectacular crimes was used to illustrate further the decision-making strategies of malefactors. The first conclusion to be drawn is that malefactors are complex, constituting a wide range of different types, with different motivations and strategies. There are many particular lessons to be learnt from this analysis, but also some more general ones. For most determined malefactors security represents a hurdle that needs to be overcome. If they cannot work out a way of defeating it they will find another means to attack the target or another target to attack. Generally malefactors have a low regard for security systems, particularly the human element. Probably the most significant factor in decision-making is the likelihood of getting caught. Thus from the security managers' perspective, in designing a security system the greater the number of layers and the more effective the strategies that are employed to increase the chances that a malefactor will be caught, the more robust the security system. This leads nicely to the next chapter, which explores what is far too often the weak link in the security system – the human element.

4
The Human Element of the Security System

> It occurred to me that it was an absurd idea of law enforcement to put an isolated, ill-paid, ill-equipped, security guard in a situation where he might have to confront a gang of seven or eight highly motivated robbers with weapons.
>
> (McLeod, 2002: 18)

Introduction

In the exploration of security failure in the previous chapter, it was shown that the human element – often the security officer – is frequently seen as the cause of the failure. Either as a result of incompetence, lack of capacity, poor reputation or plain corruption it is the action (or inaction) of the security officer that appears to precipitate the failure. The quality of the human element is effectively 'sub-prime'. However, it was also demonstrated that it is necessary to look at the broader socio-technical system to try and understand what conditions contributed to causing that human failure. If security guards are not properly trained to deal with a particular incident they cannot be blamed if they don't deal with it properly. Those broader conditions are usually the responsibility of the security manager. It therefore seems entirely appropriate to examine security officers and security managers in a bit more depth. There is a small but growing body of research in this area, from which we can begin to recognise some of their strengths and weaknesses. These will in turn offer clues for Chapter 8 in identifying strategies to enhance the human element. We begin here, therefore, by looking at the existing research on security officers before moving on to examine security managers.

Security officers

The opening quote from Ross McLeod sums up very well the problem with security officers across the globe. Below we will examine some of the issues of concern in relation to security officers, beginning with vetting, pay and conditions and continuing with training. We will also look at some of the growing body of work on security guards' occupational culture and assess how this undermines their effectiveness.

Character of some security officers

Security officers can occupy sensitive and responsible positions and it is essential that they are of good character. In many countries licensing systems are in place in order to prevent individuals with certain criminal convictions from working in the security industry. Elsewhere, however, no such licensing exists. Further, where licensing does exist individuals with less serious criminal convictions or whose convictions are spent and who have not been convicted of other offences within a specified time period are able to work in the security industry. There is much anecdotal evidence of security staff abusing their positions of trust and many exposés of criminals and/or those with links to criminal gangs operating in unregulated industries/sectors (House of Commons Home Affairs Committee, 1995). Chapter 2 presented some examples of behaviour where 'inside' staff have been corrupted into taking part in criminal acts or in facilitating them. Other evidence has illustrated the potential scale of the problem. For example the Association of Chief Police Officers (ACPO) in England and Wales published research which showed that over 2500 offences are committed annually by private security staff (ACPO, 1995). In Japan there has been evidence of security firms linked to, or even owned by criminal organisations. In 1971 of 321 companies, 20 presidents were convicted criminals, including two with links to organised crime (Miyazawa, 1991). Japanese research also presented evidence of security staff committing crimes. In 1982, 362 ordinary and 371 special penal offences were committed by security guards; of these, 55 and 92, respectively, were committed by staff while on duty (Yoshida, 1999).

Working hours and conditions of employment

The generally poor conditions of employment for security officers in the United Kingdom have been well documented (Alfredsson et al., 1991; Cumming and Winyard, 1984; Williams et al., 1984; Button and George, 1994, 1998; House of Commons Home Affairs Committee,

1995). The most significant issue relates to long working hours. The private security industry in the UK is regularly criticised for demanding 12-hour shifts and 60-hour weeks as the norm. Indeed the Chief Executive of the British Security Industry Association (BSIA) stated, 'Average hours worked in the manned guarding industry are still stubbornly 60 hours per week' (Dickinson, 2003: 3). The introduction of the Working Time Directive, which provides for the imposition of limits on working hours, has had minimal impact on the security industry as one of the first obligations of new recruits is often to sign away their rights. There have been frequent exposés of long working hours in the industry. In Button (2007a) I illustrated the example of the 'Chuckle Brothers'. These were 'floaters' who did not have a permanent site or want one, who were notorious for working 100-hour weeks. The night on which they were observed at Pleasure Southquay, they were working from 7 p.m. until 5 a.m., after which they would be driven to another site to work from 7 a.m. to 7 p.m. before returning to Pleasure Southquay for another night shift. One of them told me it was not uncommon for him to start work on a Friday night at 7 p.m. and complete at 7 a.m. on Monday morning (Button, 2007a). Security guards who sign their rights away leave the majority of them with fewer rights and protection than the industry's dogs. In 1991 a security guard who had been on duty with his dog for 28 hours non-stop was reported to the Scottish Society for the Prevention of Cruelty to Animals. The dog went home, the guard stayed on duty! (*Guardian*, 26 January 1991.)

However, it is not just in the UK that long working hours are common for security officers. In forthcoming research on South Korea the long working hours of security officers are demonstrated at three case study sites (Button and Park, forthcoming). In a residential case study all those interviewed were working over 80 hours per week. At the other two assignments, a factory and retail store, around three-quarters of those interviewed worked over 67 hours per week. Even the younger, better educated retail security officers were still working very long hours. At the time the average working week for a South Korean was 55 hours 42 minutes (Chongil, 2004). In Canada long working hours for security officers are also common with McLeod (2002) describing how he actively secured a job where he could sleep through the night.

Before the introduction of the minimum wage in the UK in 1999 there were a number of exposés of extremely low pay. Examples of guards earning £1 an hour were common and there was even an advert for a security job which stated that there was no pay, but free use of a mobile phone! (*Daily Mirror*, 1995.) Even in more reputable companies pay was generally low. The minimum wage means security officers are guaranteed at least £5.52 per hour (the minimum wage level in October 2007). Most security officers in the UK are paid at this end of the spectrum, on or just above the minimum wage. This compares to the national median weekly pay of £447 per week in 2006, which would take an employee earning the minimum wage 80 hours to earn (National Statistics, 2007)! In America low pay and conditions are also common. Research on security officers in New York found rates varying from the Federal minimum wage of $5.15 to $16.25 per hour with an average of $9.86 (Gotbaum, 2005). A national survey of security guards in 2003 found that contract guards earned $19,400 per year, which was half what a police officer earned and well below the national average (Parfomak, 2004).

Long hours lead, at the very least, to tiredness and the danger of guards falling asleep on duty. Research by Cumming and Winyard (1984: 8) found one officer typical of many who stated 'I'm like a zombie most of the time.' Chapter 8 will develop in more depth the impact of long working hours on performance. The effect of low pay is that many people see security work as only a short-term assignment, a stop-gap until a better job comes along. It also means that recruits often come from the bottom end of the labour market and may not be the best candidates. These circumstances combine to diminish the commitment and quality of the guards' work and in some cases leaves them open to bribery. They also help to shape the occupational culture of security officers, which will be considered later in this chapter.

Training

A recurrent theme in research on security officers worldwide is the poor standard of training that they receive. In many countries where there is a statutory licensing system there is also a minimum level of training that security officers must undertake. This is often the only training that security officers are ever offered. Where there are no statutory training standards it is down to companies – sometimes through the establishment of self-regulatory frameworks – to establish training standards. However, gaps in regulation in some statutory systems – most notably in relation to in-house security – mean that training standards may still be determined by employers or under self-regulatory frameworks. Therefore an assessment of statutory minimum standards only provides a partial picture – and that not generally a

Table 4.1 Training standards for unarmed security officers in selected European Union countries

Country	Statutory minimum training standards
Belgium	Security officers 120 hrs within 8 months and must pass test. Additional specialist training of 40 hrs (weapons), 75 hrs (protection of people), 16 hrs (guard dogs), 65 hrs cash-in-transit security officers, 70 hrs managers, 45 hrs (installation firm managers) and 40 hrs (installation firm staff).
Denmark	Security officers 120 hrs to commence within 2 weeks and be completed within 6 months. No exam.
England and Wales	Security officers 30 hours (22½ theory 7½ practical).
Finland	Security officers 100 hours basic training. Specialist training for managers, use of weapons and guard dogs also mandated.
Germany	Security officers 24 hours basic training and 40 hours for managers.
Hungary	320 hours training mandated.
Ireland	32 hours basic training.
Netherlands	3 weeks basic training and then additional training that leads to Basic Security Diploma which must be passed in 12 months, otherwise guard loses ID card and therefore job.
Poland	260 hours training.
Portugal	Security officers required to undergo 58 hours training and those engaged in transport of valuables or persons an additional 42 hours, those using firearms 30 hours and those in pubs/clubs 30 hours.
Spain	Security officers required to undergo 240 hours theoretical and 20 hours practical.
Sweden	Security officers 217 hrs (97 hrs classes, 120 hrs on the job). Additional specialist training of 25 hrs (security officers in public areas), 40 hrs (cash-in-transit), 40 hrs (guard dog services), 21 hrs and annual test with further 4 hrs (armed guards), and 80 hrs(bodyguards).

Notes: In Denmark courses are laid down by the police, educational authorities and the Labour Ministry also has a role; and in Sweden training is set by National Police Board. In Spain training is carried out at regional centres and in-house by accredited companies. Exam set by National Police and must be passed. Ministry of Education responsible for training of private investigators, Institute of Criminology organises courses.

good one. Let us begin, however, with the best standards in the world, which are generally to be found in Europe. Table 4.1 lists the minimum training standards of some of the countries in the European Union with the toughest regulatory and training standards.

In Austria, Cyprus, the Czech Republic, France, Greece, Italy, Luxembourg and Slovakia there are no minimum statutory training standards. Spain, Sweden, Hungary, the Netherlands, Belgium and others have good training standards. However in the three biggest economies of the European Union, Germany, the United Kingdom and France, the statutory and industry standards amount to less than a week's training.

This situation, however, is much worse in North America (Cunningham et al., 1990; Hemmens et al., 2001; O'Conner et al., 2008). Most of the states – regulation is a state responsibility in the USA and a provincial responsibility in Canada – have minimal training standards of less than a day for unarmed security officers and in many states there are no training standards whatsoever (Hemmens et al., 2001). Alaska mandates one of the highest minimum training regimes for unarmed security officers in their first year of employment of above a week (40 hours). This encompasses eight hours pre-assignment training followed by a further 40 hours within 180 days (Button and George, 2006). There is also evidence that even these low standards are frequently not complied with. Gotbaum (2005) found in New York that 12 per cent of officers interviewed had had no training, a further 17 per cent had received less than the state mandatory 8 hours, with an average of 19 hours training in 2.3 years of employment. The state law requires a further 40 hours training in the second year and only 6 per cent of officers interviewed had received even this minimum.

In Canada, research by Hyde (2003) has highlighted that with the exception of Newfoundland, mandatory training standards are a relatively recent phenomenon, with a 64-hour training course introduced for security officers in British Columbia and mandatory approved training in Saskatchewan. In Australia (where regulation is a state and territorial responsibility) in Queensland there is a minimum of 37 hours training for security officers (Prenzler et al., 1998). In New South Wales and Western Australia the National Security Competency Standards for Security Officers have been adopted, but these do not specify a minimum level of training or key subjects such as conflict resolution and the use of force (Prenzler and Sarre, 1999). In South Korea the basic mandatory training is 15 hours (Button et al., 2006). In Japan specific types of security guards are defined: airport, traffic, nuclear, valuables transport, alarm system administrator and trainers/supervisors, with qualifications set for each. However, training is only mandatory for trainers/supervisors, with the rest voluntary, and as a consequence only around 10 per cent of guards had achieved them (Yoshida, 1999).

The general picture from throughout the world is of a security officer trained for less than a week. Security officers are treated as low skilled operatives by both the state and the industry and are offered limited career structures with few opportunities for career progression. While some countries have far higher standards than these, and there are pockets of better standards within countries, these are the exception rather than the rule. For example in the UK the guarding company Wilson James prides itself on honouring the working time directive (no staff work over 48 hours per week), paying above market rates, investing in training and generally being concerned with the well-being of its officers.[1] However, it is not difficult to see that the core roles of a security officer require skills that cannot be attained in a week's training. Consider Wakefield's (2006: 388–91) functional breakdown of the roles of security officers in Table 4.2. These cover a wide range of

Table 4.2 A functional breakdown of security officer roles

Housekeeping	Prevention of losses from carelessness, fire, poor safety policies, and/or outdated or inoperative alarm systems.
Customer care	Provision of information, giving first aid, finding and reuniting lost children
Preventing crime and anti-social behaviour	Preventing theft, burglary etc.; surveillance of offenders; access control; locking and unlocking gates etc.
Enforcing rules and administering sanctions	Enforcing rules, ejecting banned people, conflict management, imposition of bans, etc.
Responding to emergencies and offences in progress	Dealing with mentally disturbed, drunk, drugged people; arresting offenders; responding to calls for assistance etc.
Gathering and sharing information	Seizing and preserving exhibits; preparing reports for police and courts; presenting evidence in court; CCTV monitoring; form filling; informal liaison with police etc.
Other	Employee education and training.

Source: Adapted from Wakefield (2006: 388–91).

functions, the majority of which most security officers would need the competence to undertake. To expect a security officer competently to engage in these skills having undertaken less than one week's training is absurd.

There is also the need to develop a career structure and to give the role a more professional orientation. There should not only be a core basic training requirement, but also more advanced training standards as well as mandatory training for specialist roles. This is an issue I will return to in Chapter 9.

Occupational culture

Only a few studies have touched upon the occupational culture of security officers. One of the first to seek to understand security officer culture was Rigakos (2002) in his study of an atypical security company operating in Toronto, Canada. Rigakos identified parallels with police culture, but also found some significant differences. Among the characteristics he identified were a culture of 'resistance from within: the art of ghosting', where security officers seek out ways of subverting the controlling culture and systems of the organisation. Second he identified a 'crime fighting and wannabe' culture where most officers wanted to become public police and revelled in their discussions of 'pinches' or involvement in crime-fighting. Third there was a culture of 'safety in numbers', in which as many officers as possible would involve themselves in incidents in the belief that this would maintain their safety. Finally he identified a culture of 'fear' in which officers were seriously concerned about the next threat to themselves. Given the unique characteristics of the security company studied by Rigakos it would be inappropriate to generalise that this broadly represents the culture of Canadian security officers. In Button (2007a) I have sought to offer an assessment of security officer culture based on two case studies in the UK. Again one must stress the caveat that the representativeness of this to the broader industry is debatable, but the structural causes of the culture are so common that I would expect at least in the sectors studied (shopping malls and factories) that it would be representative. The research identified the following traits.

1. 'Wannabe somewhere else or doing anything else'

The defining characteristic of the occupational culture of a security officer is to 'wannabe somewhere else or doing anything else', hence the cry 'I'm a security guard, get me out of here.' The research found evidence of low commitment and poor orientation to the job. There was also evidence of security officers actively seeking alternative employment, with almost half the officers at one site actively seeking other jobs despite being paid well over the minimum wage. For most security was not a career choice, but something they had fallen into. Indeed there is much other evidence to illustrate high labour turnover in this sector.

2. Challenging working conditions

Underlying the 'wannabe somewhere else or doing anything else' culture was the next characteristic, that of challenging working conditions. Common to both research sites was continuous discussion among the security officers of their working conditions. This manifested itself largely in terms of moans about the negative aspects of their conditions, but there was also a degree of bravado in being able to deal with them. These included expressions of both dissatisfaction and bravado concerning the long working hours, lack of breaks, poor facilities and the extremes of weather as well as complaints about pay. Sixty-hour weeks are the norm and these undermine the effectiveness of many security officers, who struggle to keep awake on duty.

3. Solidarity in the face of danger and isolation

Solidarity in the police and armed forces is generally seen as a positive trait and is usually shaped by the dangerous situations they face. Among the security officers a strong degree of solidarity was also found, although for slightly different reasons in each case study. In the retail facility where there were potential risks from arresting shoplifters and dealing with incidents in the night-time-economy (NTE) it was based on real feelings of danger. Only working together strongly as a team could they confront these problems. At the factory it was based upon isolation and a sense of inferiority, in that the guards united in the face of what they saw as 'them and us' – a much less positive reason.

4. Machismo

There was also a degree of machismo amongst the security officers studied, something that is frequently found in male-dominated occupations, which security is. At one level this manifested itself in views that women should not be doing certain types of security work, such as patrolling a factory at night alone – views that seem out of touch when so many far more dangerous jobs are now done by women, and not a little hypocritical given that some of the male guards who undertook this work at Armed Industries were scared of the dark themselves and didn't patrol the whole factory (Button, 2007a)! At another level it manifested itself in attitudes expressed towards women during working hours, either because of the nature of the job or because male officers tended to look at girlie magazines to pass the time. Indeed such were the delights for some officers in watching the 'eye candy' and 'totty' I was told by one officer the job gave him 'ball ache'.

5. *Suspicious and risk-focused minds*

A more positive trait among the security officers was that they became suspicious and risk-focused. Many of the officers would automatically look out for potential hazards and risks for the organisation they worked for. This ranged from identifying potential troublemakers entering the leisure facility to switching off lights and other electrical equipment left on by staff. Most were good at this, but there were a minority whose low commitment to the job stopped them developing these skills. Some guards at the factory, for instance, would choose vehicles to search because they were 'easy' rather than because there was anything genuinely suspicious about them.

6. *'Watchman parapolice' continuum*

The final trait of security officer culture depends on the orientation of the security officer on what I call the 'watchman parapolice' continuum. At one extreme of the continuum is the 'old watchman' orientation. These officers have little commitment to their role, see their job as merely to observe and report, seek to avoid confrontation and have little confidence in the quality or importance of their work. At the other extreme is the 'parapolice' orientation where there is greater commitment, a preoccupation with 'real work', and a willingness to engage in dangerous situations. These are two extremes of orientation and although many of the officers at the factory could be seen as representing the 'old watchmen' and those at the retail/leisure facility as conforming more to the 'parapolice' orientation, there were exceptions within both these groups of officers.

Clearly for many organisations it is a case of 'horses for courses' and a watchman might be what is required. However, the vast majority of organisations want something closer to the parapolice officer and in some circumstances that is exactly what is needed. At the leisure facility, where many of the officers could be described as of the parapolice orientation, officers demonstrated courage in dealing with regular drunken disorder at night and the arresting of sometimes drug-crazed shoplifters, in situations that were at times extremely dangerous. The last thing the client wanted was officers who would run away, leaving chaos and disorder in their wake and creating an image that threatened the profitability of the leisure facility.

Models of security officer

Research from the 1960s onwards has recognised the existence of different sub-cultures in relation to the police and has enabled a variety of styles and cultures to be distinguished (Reiner, 2000). Reiner (1978) identified the 'bobby', the 'uniform carrier', the 'new centurion' and the 'professional' models. There is much less research to draw upon in attempting to identify models of private security officers, but Michael (2002) distinguished four types of security officer, largely based upon their employment orientation. First she distinguished 'the casual', usually a younger security officer undertaking security on a temporary basis. Second there was the 'time server', generally an older employee engaged in security work because it tends to be non-ageist in recruitment. 'The uniformed pensioner' was the third category she defined, and it described an older security officer who had formerly been in the armed forces to whom security work was a way of supplementing a pension. Finally there was the 'police wannabe', generally a young security officer orientated towards crime control who intended to join the police. The first three types are differentiated by their employment status/orientation and it could be argued that they share some of the characteristics of the 'watchmen'-oriented end of the continuum discussed above, whereas the latter model fits the 'parapolicing' orientation. Micucci (1998) has also sought to develop typologies of security officers, dividing them between 'crime fighters', 'guards' and 'bureaucratic cops'. There are similarities between the first two and the 'parapolice' and 'watchmen' and the latter relates more to supervisors and managers. All these models are recognisable amongst security officers.

Another attempt at distinguishing models of security officers provides an interesting basis for an exploration of some of the differences between security staff. McLeod (2002) distinguished three types of security officer. The first he called 'nightwatchmen' or 'warm bodies'. These were low skilled, low status security officers – who were unable to get better paid jobs elsewhere – undertaking basic security functions. They were 'young men on their way up, old men on the way down'. The second model he called 'low profile' or 'guards with blazers'. These officers were found in more prominent locations where interaction with the public was required. They were more presentable than the first group and had more extensive security functions but they still saw the job as transitory. They were also more oriented towards observing and reporting incidents. The third model he defines as 'parapolicing' or 'private law enforcement'. These were characterised by high profile, well trained professional security staff prepared to engage in dangerous incidents and they were closer in orientation to the police.

The consequences of security officer culture

The main consequence of security officer culture is an increased chance of security failure. A few examples will further highlight how some of the negative aspects of the occupational culture lead to poor performance. In the study I published on security officers at Armed Industries (Button, 2007a) all staff were supposed to show their passes when entering the site. On gate three, which was one of the quietest and dullest posts I spent two hours with the security officer. During that period six people did not show their pass. Some of the staff just walked past, completely ignoring the security officer, others said hello but failed to show their pass. The security officer didn't do anything, claiming that he knew that they worked in the factory and held passes. He told me, 'I don't get paid enough to chase after them!' During another observation session with a female security officer I asked her on what 'random' basis she selected vehicles. She replied, 'I do a search when I feel like it. I try and pick the easy ones where you can have a quick look' (Button, 2007a). Hainmüller and Lemnitzer (2003) have also observed the link between high labour turnover, low pay and poor training and poor performance in screening at airports by security officers. They argue (2003: 4–5):

> The causal links between these variables and screening performance are straightforward. Without receiving proper training, screeners will hardly know what to look for ... A similar causal link applies to low pay. It is one of the well-proven findings of labour economics that 'you get what you pay for'. Low pay discourages highly skilled workers from applying ... The causal mechanism between turnover and performance is as follows: as with most tasks, the performance of screening increases with experience. If, as found in one study (there is high labour turnover) ... security checkpoints are rarely staffed with experienced personnel.

These few examples nonetheless demonstrate the impact of structural conditions and occupational culture on the performance of security officers. We will now consider the role of security managers, who have been the subject of even less research.

Security managers

There are dozens of books demonstrating how to undertake security management (Nalla and Morash, 2002), but very few pieces of research on managers themselves. Research in the UK, however, has suggested that less than 50 per cent of businesses in the FTSE 250 employ dedicated security management teams (McGee, 2006). This statistic was not derived from empirical evidence, rather it was based on comments by security managers interviewed in the UK. Security functions are undertaken in these organisations, but they are the responsibility of another department, such as facilities management. Nonetheless many organisations do employ security managers in managerial or lead advisory roles on security. They have a multitude of different job titles. Among the most popular are security manager, security management specialist, security adviser, security controller, security coordinator, security administrator, chief security officer, security director, security analyst, security specialist, loss prevention manager, director of surveillance and protection manager. These may also be preceded by a range of titles such as assistant, group, senior, divisional and deputy. Security managers (which will be the generic term used here) have a wide range of duties. Nalla and Morash (2002) have assessed textbooks on security management to provide a list of the full range of functions generally undertaken by security managers, and have further refined the information to arrive at what they consider to be some of the main functions of a security manager. Table 4.3 illustrates these.

The functions in Table 4.3 will be familiar to most security managers, but some will recognise functions that are not their responsibility and others will see gaps – such is the lack of common ground in what security managers do. Some security managers have no responsibility for certain security functions. For instance, information technology security is often hived off to a specialist in the IT department or security IT department. In some organisations crisis management is the responsibility of a specialist crisis manager, contingency planner or some other manager. A number of organisations may also have a separate investigations unit or a counter-fraud department.

Some security managers also have additional non-security related functions. Hearnden (1995) found that the most common additional function undertaken by security managers was fire prevention and

Table 4.3   Nalla and Morash's functions of a security manager

Function	Sub-functions
Personnel protection	Executive protection Employee security awareness programme Other employee protection Workplace violence and response Executive protection, overseas travel Non-executive travel protection
Access control	Tracking security threats Forecasting threats Physical protection Security systems design Alarm monitoring Guard force management
Asset protection	Fraud prevention Protecting competitive data Protecting trade secrets Information security systems Patent enforcement Inventory control
Investigations	General investigations Internal investigations Investigation of crimes concerning company Fraud detection Employee background investigations
Risk management	Crisis management Disaster preparedness Emergency preparedness Due diligence Business resumption planning Promoting employee job satisfaction Promoting employee commitment Business impact analysis
Other security functions	Prevention of/response to substance abuse Protecting company goodwill Strategic planning Prevention of/response to sexual harassment

Source: Adapted from Nalla and Morash (2002: 14).

response. In earlier research he identified training, customer liaison, safety and recruitment as other common responsibilities (Hearnden, 1993). Perhaps the main variation in function between security managers in different organisations is the extent to which they have a responsibility for the management of security or whether their job is more focused on advising general managers on the use of security. In most instances the security manager is likely to have a mixture of direct responsibility and a more advisory role (George and Button, 2000).

Differences are also evident in relation to whether they report to a senior director or some other manager below board level. Research by Hearnden (1995) found that 58 per cent of those security managers surveyed reported to a director or to the main board. More recent research on American security managers in Fortune 1000 companies found that nearly three-quarters reported to either the chair, president, senior vice president or vice president of the board (Nalla and Morash, 2002). Given the growing resources and attention that many organisations have devoted to security post-9/11 it is likely that the level of reporting to UK boards has also increased (Briggs and Edwards, 2006).

In contrast to the situation of security officers some managers are paid very well. A survey for the Security Institute (TSI) found top of the range 'Chief Security Officers' in large multinational blue chip companies handling budgets over £30m and earning £125k to £150k per year plus benefits (pension, car, health insurance, shares and so on). A national security manager, responsible for a company in one country, with a budget of £2m to £10m, has a median salary in the range of £55k to £65k plus benefits. For the HQ or site manager the median is £50k to £55k (The Security Institute, 2007). The TSI survey probably overestimates the pay of security managers. Those working in the public sector – in hospitals, for local authorities and so on – would be unlikely to enjoy similar salaries. Nevertheless for many security managers the pay is good and for a substantial number, this supplements a pension from a previous career.

There is a common perception that security specialists are largely former police officers or ex-servicemen. According to research carried out by Hearnden (1993) this was an accurate view, but the percentage was declining. It seems likely that if the same research was undertaken today the percentages would be lower still. In 1989, no fewer than 86 per cent of security managers were recruited from a military or police background. In 1991, this had declined to 76 per cent and it was 61 per cent by 1993. More recent research on North American security managers found that three-quarters came from a traditional background with 31 per cent from the police, 19 per cent from intelligence and 21 per cent from the armed forces (The Conference Board Survey cited in Briggs and Edwards, 2006: 78).

Given that security management is often a second career it is no surprise to discover the average age found by Hearnden (1995) was 50.2 years. Almost three security managers in five had left school at the age of seventeen but on average they had achieved higher than the national average at GCE 'O' level and about average at GCE 'A' level. He also found that security managers worked on average fifty hours per week and earned on average £25,796 per annum (Hearnden, 1993). More recent research by Jones and Newburn (1998) found that 17 per cent of firms in the security industry employed former police officers, although this covered all ranks and not just more senior personnel. Hearnden (1993) also discovered some negative orientations concerning training and education. Of the surveyed security managers he found:

- 62 per cent had no vocational qualifications and managers rated the possession of them as the fourth least important attribute of a good security manager.
- 38 per cent had not attended at least one outside course or seminar in the previous two years.
- 59 per cent worked for organisations which had no formal training needs analysis.
- 40 per cent were unable or unwilling to identify a single personal training requirement.

It should be stressed that this research is dated and that since it was published there has been an expansion in higher education related to security. Nevertheless there are still many working in the industry as a second career after the military or police to supplement their pensions. In many areas this is self-perpetuating, as ex-military and ex-police personnel appoint subordinates and successors from the same background as themselves.

There are some negative elements to this second career mentality. It may influence many managers' orientation, as the job is supplementary in financial terms and is something that many have achieved through experience rather than qualifications. This is unlikely to promote much in the way of incentives or desire for further training and education. When security management is juxtaposed against other managerial specialisms, such as personnel, safety, risk and so on, these others are dominated by people who have made it a first choice career, who are in it for the long haul and as a consequence are prepared to invest time in securing the appropriate development through training and education to achieve their position. It is also valid to question the relevance of the experience brought to the job by some former police and military personnel (Manunta, 1996). Manunta (1996: 235) argues,

> Those employed in the private sector tend to be mature, retired people with military or police background who are unlikely to have had education at university level. Most of them have little or no career prospects, and some on their own admission are in search of a 'warm, comfortable retreat'.

In some nodes the security manager needs to be attuned to the business environment and to be able to talk that language. There is a view amongst many in the security community that some security managers drawn from the ex-services/law enforcement community cannot talk business. For example in Gill et al.'s (2007: 51) review of the value of security one interviewee stated,

> From police and military I have seen a few who are good and there has been a missing of significant opportunities for business. Especially the ex military, they are like kids with no organisational awareness. Their people skills and ability to understand cultures of business are lacking because they have not grown up in a business environment. In something like nuclear work then a military background may be important. The greater the competitive environment in which a company participates the less easy it is to appoint someone from a military background. There needs to be the most effective cultural fit.

The mentality of many police is more reactive and more focused upon dealing with incidents once they have happened than on prevention (Johnston and Shearing, 2003). Indeed many studies have documented the lack of interest in prevention amongst the police (Weatheritt, 1986; Gilling, 1997). As a consequence former police who find themselves in the more forward looking security world might find their experience of little value in developing a holistic preventative strategy. As Challinger (2006: 586) argues,

> Some security decisions appear to be made after a breach of security has occurred. Some are made when it is simplistically assumed that continued security is no longer needed. Some are made when it is

feared that business will be lost if security is not in place to reassure customers.

Challinger also goes on to argue that many security decisions are not based upon evidence. In better-established professions, such as medicine, a doctor will prescribe treatment or advice on preventative measures that are based upon scientific evidence and is also able to keep up to date on the latest advances through reading appropriate journals and attending conferences. In some of the newer and more comparable professions, such as personnel management, there is also a much stronger evidence-based approach in which selection procedures are likely to be informed by the latest thinking in professional journals and attendance at conferences (McGee, 2006). Security, in contrast, often relies on techniques that are well established and are believed to work, but for whose effectiveness there is little actual evidence. Indeed in one visit to a head of security at a factory I was shown a security manual published in 1969 and told that this was the 'bible'. The manual was good, but methods of security have advanced significantly in many areas since 1969.

Perhaps another good illustration of the downside of the mentality of many security managers is their influence in the boardroom. As Garcia (2006: 513–14) argues:

A common theme of customers and security professionals alike is that the business case for security must be made in order to acquire the resources necessary to protect assets. It is agreed that this is a necessary step, but there appears to be a lack of preparation by many security professionals in making this case, particularly compared to their peers in other divisions across the enterprise.

Gill et al. (2007) have noted the inadequacies of many security managers from an ex-services and law enforcement background in embracing and articulating the language of business. In one illustrative quote a security manager argued, '...most senior security people are just plain thick. Many cannot write basic policy or process, as much as they may understand what needs to be achieved and they cannot articulate a business case' (Gill et al., 2007: 52). There are many challenges to getting security taken seriously in the boardroom. To many boards security is not considered a priority and neither is it integrated into the broader strategies of the business (Challinger, 2006). Security may not be seen as relevant because it is wrongly believed that staff theft, fraud or comparable incidents do not constitute a problem. Some boards actually see security as a nuisance, something that gets in the way of the core business and creates additional bureaucracy. Worse, some boards might even consider security as the enemy – harassing 'honest' staff. Therefore a security manager needs to be able to make the case for security to board managers by talking their language and fitting in with their agendas. Unfortunately within many organisations there is a belief that anyone can do security (Challinger, 2006).

From their research on the value of security Gill et al. (2007) were able to identify two models of security manager (ideal types). The main characteristics of the model are set out in Table 4.4. The 'traditionalists' are associated with ex-military and police personnel, while the 'modern entrepreneurs' tend to be from more conventional business backgrounds. It is, however, important to add a caveat to the model that not all ex-military and police fit the former category and that not all those from business backgrounds fit the latter. I have met security managers from the police and military who would fit into the latter category, as well as 'traditionalists' with no police or military experience. The model, however, does provide a useful basis for debate on the kind of security manager required. There are clearly parallels to the debates in the 1980s over personnel management, with many advocating and embracing human resource management as a more appropriate model in which the functions of personnel are more closely aligned to the business objectives of an organisation (McGee, 2006). The clear drift of Gill et al.'s report is that more security managers should become 'modern entrepreneurs' and while I would agree with the broad thrust of that, for some areas of the public sector the commercial

Table 4.4   Gill et al.'s model of security managers

Traditionalists	Modern entrepreneurs
Security is a service function	Security part of business process
Necessary cost on bottom line	Security integral to all activities
Experience of police and military important in running security	Importance of influencing people and policies
Organised by command and control	Emphasis on change management
Success measured in arrests	Importance of objectives, strategy, measurement, ROI and impact on bottom line.
	Business skills more important than security expertise

Source: Gill et al. (2007).

skills of the 'modern entrepreneurs' are less crucial. In nuclear facilities, high profile government locations and military facilities, for example, aspirations to 'absolute security' override the 'bottom line'.

It is important to note here that I am not arguing that former members of the police, the military and other comparable occupations should be banned from the private security industry (although restrictions do exist in some countries). They can bring valuable experience and be part of networks that are very useful to enhancing security (Briggs and Edwards, 2006). Rather those who become security managers should also have undertaken appropriate professional training and or academic study in security. This is an issue that will be developed in more depth in Chapter 9.

Conclusion

Chapter 2 demonstrated that the human element is a common factor in causing or contributing to security failures. This chapter has explored that human element in greater depth and has suggested some reasons as to why the human element does frequently fail. It began with an exploration of the literature on security officers, outlining problems in the character of some, the poor pay and conditions, the often inadequate training and the impact that this has on the occupational culture. The following discussion on security managers showed the wide range of roles that they undertake, and examined some of the limited research on their backgrounds, which affirms a dominance of former military and police personnel. With security managers the problems included lack of interest in training and education, limited influence in the organisation and an inadequate professional approach. The chapter ended with the recent identification of two models of security management identified by Gill et al. (2007) – the 'traditionalists' and the 'modern entrepreneurs'. In short, what has been revealed here is the 'sub-prime' quality of security staff at both guard and managerial level. For security systems to be more effective the quality of the human element needs to be addressed. Chapters 8 and 9 will set out an agenda to achieve this, but it is also important to note the negative impact of some of the foundations of security on the quality of the human element, which will, among other issues, be the subject of the next chapter.

5
The Foundations of Security

> ...but where you have no organization, no parity of bargaining, the good employer is undercut by the bad and the bad employer is undercut by the worst.
>
> (Winston Churchill, cited in Dickens et al.,1994)

Introduction

The previous chapters have illustrated at a base nodal level some of the weaknesses that contribute to poor security. In this chapter we move to consider some of the more macro level weaknesses which ultimately provide the foundations upon which much of the failure and success of security are built. The first area this chapter will consider is the inadequate regulatory framework that exists in the UK and in many other countries. The nature of security requires effective statutory regulation and it will be argued that in the UK, and in many other countries, there is a lack of such regulation, which ultimately contributes towards poor security. The second area explored in this chapter is the lack of a professional infrastructure for security. For security to be more effective, the 'traits' of a profession are required and this chapter will argue that in the context of the UK – as in many other countries – this is lacking. Finally the chapter looks at the emerging security inequity in society and argues that this also needs to be tackled, not just for the benefit of the less protected nodes, but for the benefit of all.

Inadequate regulatory framework

Central to the foundations of good security is an effective system of statutory regulation based upon licensing. Regulation in some form exists in almost every country (Button, 2007b; Hakala, 2007). In the vast majority of cases, however, it does not constitute an effective system and there are only a handful of examples of good practice (Button, 2007b; Button and George, 2006). Before some of the problems with regulation are explored it is worth noting why regulation is required.

The underlying rationale for the regulation of security is that poor security has implications for public safety and that unregulated market mechanisms cannot be relied upon to ensure a minimum standard that would guarantee public safety. As Churchill observed in the quotation cited above, the consequences of no intervention are a downward spiral. Therefore minimum standards, enabling competition to take place, but without falling below the safety net, are essential. The debate over regulation has focused most significantly on the human element. Clients of security, who entrust officers with the protection of life, property, reputation and the like want security staff to be of good character and – as far as possible – of unblemished record. This is because most security positions provide major opportunities for those who are tempted to conduct crimes. Ensuring minimum standards of character is very difficult, if not impossible, to achieve without statutory licensing. Calls for controls have also centred on more general standards of quality of performance as well as improving the accountability of security staff (Button and George, 2006). As Chapter 2 showed, there have been many instances in which poor quality security contributed to significant failure, most notably on 9/11.

It is worth revisiting the nature of security in an unregulated market as it existed in the UK before 2001. For many purchasers, security is a grudge cost and they will invariably choose the cheapest option (George and Button, 2000). This leads to continuous undercutting of one security firm by another and to purchasers frequently changing contractors. Reductions in costs can only be achieved by strategies such as lowering pay, and minimising or removing training and staff vetting. The consequences of this were catalogued in frequent exposés of criminals working in the industry, incompetent acts by guards and poor standards (see House of Commons Home Affairs Committee, 1995; George and Button, 2000). Statutory regulation was seen as a way of setting minimum standards to ameliorate these problems and above which competition could take place.

In 2001, after 30 years of campaigns for regulation, the Private Security Industry Act (PSIA) was finally passed in England and Wales (it has subsequently been opted into by Scotland and Northern Ireland).

The legislation established a new non-departmental public body (a quasi-autonomous non-governmental organisation, or QUANGO) called the Security Industry Authority (SIA) (Button, 2003). The SIA has a wide range of functions, but the most important relate to the licensing of individuals operating in sectors in the private security industry subject to regulation. As the legislation stands this includes:

- Door supervisors (contract and in-house)
- Vehicle immobilisers (contract and in-house)
- Security officers (contract)
- Keyholders (contract)
- Private investigators (contract) (not yet implemented)
- Security consultants (contract) (not yet implemented)

Licensing is subject to an identity check, a criminal records check and a competency requirement. The standards vary for different types of licences. In designated sectors licences are required for all employees from the most junior to the most senior. The legislation did not introduce the compulsory licensing of firms, but instead set out a voluntary scheme, although again there is scope to make it compulsory under secondary legislation.

The legislation gave a great deal of discretion to the Home Secretary, the chief executive and the chair (and board) of the SIA in creating the standards and in the capacity to create secondary regulation. The first appointments to the positions of chair and chief executive were therefore to be crucial to the direction of the SIA. The first chair was Molly Meacher, a former board member of the Police Complaints Authority, and the first chief executive was John Saunders, a former banker. Initially the SIA seemed to be considering the pursuit of a much tougher regulatory agenda. Speaking at a conference in October 2002 Molly Meacher said

> We are very interested in Europe. Most interesting to us is Sweden. That's why we are going there later this month, because they seem to have quite a radical regulatory system similar to what we are trying to do in Britain. We are trying to learn how others have gone through the problems we are currently struggling with.
>
> (Professional Security, 2002)

Little over a year later she had resigned, offering no comment on the reasons why. Speculation in the industry, however, suggested differences of opinion with the chief executive over the ambitions of regulation. Indeed as the following section and comparison with other sectors will show, the system that has emerged in England and Wales falls far short of Molly Meacher's Swedish ambitions and owes more to the minimalism of American standards of regulation.

The first concern relates to the ambition of some of the standards created. For security guards the SIA has mandated 30 hours of training consisting of 22.5 hours of knowledge-based training and 7.5 hours of practical-based training. Under the old voluntary system, which used British Standard (BS) 7499 (and which was followed by nearly all the large and medium-sized companies) the mandatory training was two days' knowledge-based and one day practical. Essentially, then, regulation has resulted in one extra day of training on conflict management and communication skills. Any increase in mandatory training of security officers is to be welcomed, but this falls far short of Sweden's 217 hours, encompassing 97 hours of theoretical and 120 hours practical.

Another concern with the training system that has been introduced under regulation is the decision to allow training companies themselves to determine whether a candidate has passed the minimum training. Evidence has emerged of training companies being very slack in the administration of exams for what are already very basic standards (allowing cheating, giving hints for answers etc.) ('Panorama', 2008). It is as though driving instructors were to administer the driving test rather than an independent government agency, a situation that would not be tolerated as it would undoubtedly lead to many candidates being passed as fit to drive when they should not. This is what is happening with security officer training and it exacerbates already low standards.

An illustration of the lack of ambition of regulation and perhaps one of its most significant weaknesses is the 'Approved Contractors Scheme'. First of all it is voluntary, so in the cut-throat security market there is scope for some firms to undercut others by avoiding the cost of the standards that approved contractors must meet. More importantly, however, there is no proper sanction for firms that flout regulations. For example, if an 'approved contractor' regularly breached employee regulations the firm might lose approved status, but could still continue trading. Worse still, if it was not an approved contractor, there would be no sanctions at all. In most regulated industries breaches of regulation can lead to penalties and even revocation of the firm's licence. For example, in the financial services sector, which is regulated by the Financial Services Authority (FSA), Norwich Union was fined a record £1.26 million in 2007 over failure to protect customers' personal information and thus putting them at increased risk of fraud. The FSA found that fraudsters were able to get hold of personal information on 632 policy holders. Norwich Union agreed to settle early in the investigation and as a consequence the fine was reduced by 30 per cent (Timesonline, 2007). This kind of sanction, with the ultimate threat of withdrawal of a firm's licence for gross breaches in regulation, is the most effective means of ensuring compliance.

Finally, the approved contractors scheme is largely based on existing British Standards and accreditation through United Kingdom Accreditation Service (UKAS) approved inspectorates, such as the National Security Inspectorate (NSI). Firms registered with these are entitled to a fast-track accreditation through the SIA (SIA, 2006). This scheme gave the SIA an opportunity to create a much more ambitious set of standards that would have helped to raise the standards throughout the industry, but instead it has maintained the status quo and a set of standards that will do little to improve the overall performance of the industry. The Approved Contractors Scheme should be abolished and replaced with compulsory licensing of firms.

The failure to regulate in-house security guards is another significant omission. This means that a large number of security officers are not required to have a licence, which undermines the aim of improving the industry and will lead to confusion amongst the public and other agencies. There is the strong likelihood of in-house and contract security officers working alongside one another in places like shopping malls; with contract officers trained, licensed and subject to the governance of the SIA, and in-house officers beyond the scope of the SIA and meeting only those standards imposed by their employers. With the high cost of licences (£245 or $490) (for three years) and training (£500+ or $1000+) some organizations may well be tempted to employ cheaper in-house officers and avoid regulation.

The costs have also been criticised by some security personnel as excessive, which seems fair when the costs of licensing in the UK are compared to costs in the Republic of Ireland where regulation was introduced not long after England and Wales with the establishment of the Private Security Authority in 2004. Here the cost of a security guard licence is only €80 (£61) (for two years) (Private Security Authority, 2007).

As well as omissions such as that relating to the in-house security sector, the PSIA system also neglected elements of the installation sectors relating to locks, intruder alarms and so on. Additionally the

SIA has been slow to implement regulations in regard to private investigators and has deferred regulating security consultants, something they have the power to do. The gaps in the legislation and their own decisions against licensing have left loopholes in the SIA's system open to exploitation by criminals. For example, a 'Panorama' investigation into regulation found former criminals – with histories of serious convictions – running security companies and evading regulation through becoming 'security consultants' (Panorama, 2008). Gaps in legislation will always be exploited by the unscrupulous, which is why all of the industry should be subject to licensing.

Another issue concerning the effectiveness of the legislation concerns the emerging mentality of regulation. The SIA has initiated regulation based on collaboration and working in partnership with the industry, customers and public (SIA, 2004). This is similar to the 'responsive' regulation advocated by Ayres and Braithwaite (1992), which is characterised by the interplay of self-regulation and state regulation, and attempts to integrate those who are subject to regulation and other interest groups within the regulatory process. Such a model is clearly preferable to over-zealous state regulation, which undermines the autonomy of the regulated, or to pure self-regulation, which usually has little impact. Nevertheless there is a fine line between 'responsive' regulation and 'capture', where the regulator starts to act in the interest of the regulated, rather than in the 'public interest'. At a general level the SIA has been criticised by Zedner (2006) as becoming the 'pimp' of the industry. More specifically, murmurs of discontent over the SIA's relationship with the largest trade association (representing security firms), the British Security Industry Association (BSIA), have begun to emerge (Frosdick, 2005). The dissatisfaction has focused upon the SIA's reliance on the BSIA as the main consultative stakeholder. The British private security industry is highly fragmented with dozens of representative associations and many have felt left out in the cold.

Ultimately it is too early to make a case for 'capture' – and, to be fair to the SIA, the fragmentation of the British industry does make consultation very difficult (George and Button, 2000). The relationship between the two organizations, however, does require careful monitoring. What has emerged also offers broader evidence of the ability of some nodal assemblages to exploit legal space for their own benefit. Clearly the BSIA and the larger security companies have been much more adept at this than other interests. We now turn to regulatory experience in the European Union, as this offers further insights into both poor and very good examples of regulation.

Regulation in the European Union

Most countries in the European Union have some form of regulation; the Czech Republic and Cyprus are the only exceptions. There is, however, wide variation in the regulatory systems for the private security industry. The least sophisticated system is Luxembourg's, which centres on the requirement for a licence for a security firm from the Ministry of Justice. There is no licensing of employees, minimum training requirement or character standards that would bar persons from working in the industry. There are requirements relating to the carrying of identity cards, and collective agreements between the firms and trade unions set minimum working hours and pay (Weber, 2002). At the other extreme is the Spanish regulatory system wherein both employees and firms are required to possess a licence and there are standards of character and training for personnel. The training standard for security officers includes 240 hours theoretical and 20 hours practical, with refresher training of 75 hours mandated every three years. Standards also apply on uniforms, weapons, guard dogs and the financial resources of the undertaking. Training standards also apply to managers and, uniquely, in-house security is prohibited, so organizations wishing to employ the equivalent to in-house must set up their own security firm. The law also sets out the activities that the private sector may undertake. Such is the ambition of the Spanish system that it has sought to integrate the private sector in the public justice system as a complementary arm to the state security infrastructure (Gimenez-Salinas, 2004).

Another country with a demanding regulatory system is Belgium, where the establishment of a private security undertaking requires a licence from the Ministry of the Interior (in consultation with the Ministry of Justice) and there are stringent character requirements for owners, managers and staff. There are also restrictions on the activities of owners/managers of security firms, as they cannot perform private investigatory, arms' dealer or any other function that could endanger public safety (Weber, 2002). Standards also exist for uniforms, vehicles and weapons, and require the submission of annual reports. Training standards are high, with a mandatory number of hours training for particular sectors/activities. Basic security officer training is 130 hours, training for protection of persons 66 hours, transport of valuables 78 hours, middle managers 40 hours and top managers 106 hours (CoESS/UNI Europa, 2004).

Button (2007b) undertook a country comparison of regulation in the EU-15. To facilitate the creation of a league table it was decided to focus purely upon the regulatory system in operation in each country for general unarmed security officers in the static manned guarding sector, where more objective analysis could be pursued. The study focused on performance issues and did not assess governance issues such as which body was responsible for regulation, the nature of complaints procedures and so on. The other major omission was the enforcement of the regulatory system and the level of compliance with regulations. This is clearly an important criterion, but an assessment of enforcement and compliance would require subjective judgements based upon data that would probably be of varying quality from different countries and which was not available across the countries studied.

The criteria for the league table were influenced by the 'Joint Opinion of the European Social Partners in the Private Security Industry on Regulation and Licensing' which was signed by CoESS (Confederation of European Security Services) and EURO-FIET (the predecessor to UNI-Europa) in 1996. The declaration set out the following principles vis-à-vis regulation of private security:

- Effective regulation of private security is essential for high standards of professionalism and standards in the private security industry.
- All firms in the private security industry should be licensed.
- Employees should be screened and those with certain criminal convictions should be barred from employment.
- Employees should undertake vocational training.

In light of these principles for the analysis of different regulatory systems the first criterion was whether there is a licensing system (or equivalent) for security staff. Thus if there was a system that enabled those with undesirable character to be refused permission to work in the security industry it was classed as a 'yes' and the country was given two points and if not it received no points. The next assessment was based on whether there was any security-specific licensing regulation for firms and if yes, two points were awarded and if no, the country received no points. The next criterion related to the mandatory training required for general security officers. As a quality enhancing aspect of regulation it has frequently been identified as one of the most important criteria (APEX, 1991; Spaninks et al., 1999; CoESS, 2004). It was therefore decided that the points available would be doubled to four for training of security guards. Mandatory training does vary significantly across the EU, so the scale ranged from zero where there was no mandatory training, to one point for up to 40 hours, two points for 41 to 80 hours, three points for 81 to 120 hours, and four points for 121 hours or more. Linked to training for security officers it is also important that standards exist for those who manage them. Therefore if there were any minimum experience, training or qualifications for managers, two points were given and if there were none, no points were given. These two criteria together provided half the available points, which was appropriate given the importance of training in enhancing the quality of security provision.

A common gap in regulatory systems is the (non-)regulation of in-house (or proprietary) security officers (National Advisory Committee on Criminal Justice Standards and Goals, 1976; De Waard, and Van De Hoek, 1991; Police Foundation and Policy Studies Institute Independent Inquiry, 1996; Prenzler and Sarre, 1999; Hyde, 2003; Hakala, 2007). This gap undermines the effectiveness of a regulatory system in two ways: first, by providing a loophole for those seeking cheaper alternatives to regulated staff; and second, by having security staff who may not reach the same standards as regulated staff. Therefore it was decided to add the final criterion of whether in-house security guards were subjected to regulation. If yes, they were given two points and if no, they received no points. It was also decided to focus upon the 15 EU members before the most recent enlargement. Except for Cyprus and Malta, which are very small countries, the most recent entrants are former communist countries with relatively young security industries and regulatory systems. Nevertheless some, for example, Hungary, have already introduced very ambitious regulatory systems (CoESS/UNI Europa, 2004).

Table 5.1 shows that two countries achieved the maximum 12 points: Spain and Belgium. As they had the same number of points the order was decided according to training standards: Spain had the highest with 260 hours, compared to 130 hours for Belgium. The Netherlands was next with 11 points, although its training standard of two weeks followed by a requirement to secure a diploma within a year meant that under the system used for this paper it was at a disadvantage to the norm of countries mandating hours, although the reality might be a better training package. Portugal and Sweden followed on 10 points each. Portugal was let down by a training standard of only 58 hours for security officers, and Sweden lost ground because it does not regulate in-house security, although it does have mandatory training of 217 hours for security officers. Denmark, Finland, France and Ireland also have good systems with 9 points each.

Table 5.1   League table of the regulatory systems for the static manned guarding sector in 15 EU states

State	Licensing of contract security guards	Special licensing of security firms	Mandatory training for security guards	Managerial competence/ training standards	Licensing of in-house security guards	Total points
Spain	2	2	4	2	2	12
Belgium	2	2	4	2	2	12
Netherlands	2	2	3	2	2	11
Portugal	2	2	2	2	2	10
Sweden	2	2	4	2	0	10
Denmark	2	2	3	2	0	9
Finland	2	2	3	2	0	9
France	2	2	1	2	2	9
Ireland	2	2	1	2	2	9
Austria	2	2	0	2	0	6
Germany	2	0	1	2	0	5
Greece	2	0	0	0	2	4
Italy	2	2	0	0	0	4
England and Wales	2	0	1	0	0	3
Luxembourg	0	2	0	0	0	2

Source: Button (2007: 122).

At the bottom of the table is Luxembourg with a very basic system. However, with such a small country and industry it could be argued that it is much easier to control through self-regulation, collective agreements and peer pressure. Regulation in England and Wales, with one of the largest industries in Europe, does not have this get-out clause. Minimal basic training provision, failure to regulate in-house security, a voluntary approved companies scheme and a lack of training/competence standards for managers means that in reality it is at the bottom of the table. Austria, Greece and Italy all lack mandatory training standards and have other significant gaps in their legislation. While as noted earlier, the best regulatory system may not necessarily mean the best security industry (this is a topic for further research), the regulatory system is a major influence on the quality of the industry and it would seem only common sense that a security officer who has undergone 200+ hours training is generally going to be a higher quality operative than one who has had less than a week's training. The bottom line is that in the EU-15 only the top five in the table, with 10 points and more, have systems that set the standard of what is required in regulation. Much of the EU exceeds what is set elsewhere, so for regulation to begin to enhance the quality of private security across the world, much needs to be done. Chapter 9 will explore some of the principles of regulation that need to be established to achieve this.

Restrictive regulation

At the other end of the scale there are also some regulatory systems that frustrate the private security industry with excessive and overbearing regulations. Usually these are systems that have been created by those sharing the 'radical negative' and 'conservative negative' perspectives. For example a report on private security regulation in South-East Europe illustrated a wide range of regulation that was clearly designed to frustrate the private security industry. In Albania a security company cannot employ in any district more than five per cent of the number of police employed there (International Alert, 2005). Nevertheless in the context of Albania – and perhaps well-founded fears of corrupt security companies running a locality rather than the state – such regulations might be entirely appropriate. In Romania legislation recently sought to mandate the same uniform for every security officer. This was subsequently withdrawn after an outcry from the security industry. In many US states standards of character and training are eclipsed by extensive requirements on uniforms, what words must be shown on badges, vehicle markings, ID cards and so on. Several states even go as far as mandating appearance standards for security officers. For instance Alaska mandates several conditions relating to appearance.

> Natural hair must be clean, neat and combed. Hair must not extend below the top of the shirt collar at the back of the neck when standing with the head in normal posture. The bulk of the rest of the hair must not interfere with the normal wearing of all standard headgear.
>
> (Alaska Statutes Section 13 AAC 65.010)

While it is entirely appropriate that a security guard should be orderly in appearance, whether that should be mandated in regulations is more debatable. The regulatory framework should shape the broad standards and avoid becoming too tightly embroiled in detail that becomes difficult to enforce. In Chapter 9 the principles of an effective regulatory system will be outlined. The lack of effective regulation is linked to the next element of the poor foundations, the lack of a professional infrastructure.

Lack of professional infrastructure

There has been a long-running debate over whether security management can be considered a profession (Manunta, 1996; Simonsen, 1996; McGee, 2006). The mere fact of the debate might suggest that security management is not a profession, for such debates do not occur in established professions, such as the law, surveying or dentistry. The definition and characteristics of professions have been the subject of a huge amount of research – and indeed debate – amongst 'trait theorists', who have sought to identify the key elements of a profession and measure occupational structures against these; 'functionalists' who focus upon the aspects of professional life; and finally the 'interactionists', who centre upon the institutional self-interest demonstrated by such groups (Millerson, 1964; Parsons, 1951; Larson, 1977). Given that there is insufficient space here to explore all these approaches in depth, we will focus on the 'trait' approach to consider at a base level to what extent security management maps as a 'profession'. If we consider some of the main 'traits' and other characteristics of a profession, security management in the UK reveals a mixed bag of positives and negatives.

Before we embark upon this, however, the question must be asked: does security management need to be a profession? There are many occupations which do not aspire to become professions, so why should security management? The answer lies in the importance that security management has for many parts of our lives. It has fundamental responsibilities relating to the protection of life, property, financial viability and reputation, to name only a few. To maximise this protection requires people who can utilise expert knowledge, and this is most effectively achieved through a professional model where there is a body of knowledge that security managers know and update as it changes, and where functionaries have to pass tests on that knowledge to practice and are governed by norms relating to their conduct and ethics. This is why security management needs to be a profession. However, let us first consider the evidence on the progress towards a profession through a number of traits.

1. A body of knowledge

Security – as this book will show – has a huge body of knowledge to build on. The problem, however, is that it is largely based in other disciplines and access to it can therefore be difficult for security practitioners. For example, there are bodies of research in – among other disciplines – criminology, psychology, geography, planning, design, computer science and risk management that would be of use to the security practitioner. There is also, however, a paucity of outlets to disseminate such knowledge to security managers and, even if such outlets existed, the attitude of many security managers to such publications is broadly negative. Only a handful of books, and academic journals such as *Security Journal*, and academics such as Professor Martin Gill have sought to bridge this gap. Although matters are improving there are still underlying mentalities in security management that hamper this. The fact is that many security managers could probably not even name an academic journal, let alone claim to have read an article or to do so on a regular basis. There is no culture of evidence-based practice where solutions are identified based upon knowledge of what works. As Gill (2007: 8) notes,

> ...it would be difficult to argue that current security practice generally, as well as retail security specifically, is predicated on a body of knowledge derived from scientific enquiry. The fact of the matter is that few studies have been undertaken and what has been done is not systematically made available to decision-makers at the sharp end of practice.

Compare this to more established professions such as medicine. Here a doctor prescribes a treatment to a patient based upon research that has been taught to him/her in medical school, at a conference or training event, or identified in a professional journal. Such a mode of working is not the norm in the security world.

2. A recognised professional association

A second trait that is frequently cited as characteristic of a profession is a single professional association that covers most, if not all, of those it seeks to represent. Such bodies should not only represent the occupation but identify standards and provide training, education and development opportunities. So, for example, with the case of medical doctors in the UK there is the British Medical Association and for solicitors, the Law Society. In the British security industry, as in many

Table 5.2 Professional associations for general security managers in the UK

Name of association	Profile	Competence entrance requirements	Membership
The Security Institute (TSI) http://www.security-institute.org/. As of January 2008 it merged with the International Institute of Security (IIS).	UK based organisation founded in 1999 to promoted professionalism amongst security managers. Has many of the most senior security managers and is most influential. Undertakes seminars.	Accreditation	1000+ (now IIS merged)
	IIS was UK-based organisation founded in 1968 with only exam-based entrance in UK. Has wide base that extends beyond the UK, but has been in decline since mid-1990s. Undertakes seminars, provision of newsletters.	IIS was exam or relevant qualification	
American Society for Industrial Security (ASIS) http://www.asis.org.uk/.	American-based organisation with a worldwide membership base and significant UK chapter. Has many senior security managers as members and is also influential. Undertakes seminars, provision of newsletters.	Fee.	35,000 worldwide, 750 in UK.
Institute of Security Management http://www.i-s-m.co.uk/.	UK-based organisation largely representing middle ranking security managers based in Midlands. Undertakes seminars and has newsletter.	Fee.	Unknown
International Security Management Association http://isma.com/.	US-based organisation with worldwide membership focused upon larger blue-chip companies.	Role, size of organisation, degree or experience	Unknown

other countries, the situation is generally much more fragmented (George and Button, 2000). As Table 5.2 shows there are four associations alone covering general security management. This also fails to take account of the even more sector-specific security manager organisations covering defence, universities, health care and so on. Consolidation is essential if a qualitatively enhanced, coordinated and influential profession is to emerge. IIS and TSI have recently merged, which is welcome, but much greater merging and consolidation is required. The new organisation also needs to pursue a much stronger agenda of professionalisation, something which will be explored in Chapter 9.

3. Training, education and entrance requirements

Chapter 4 demonstrated the significant influence of the military and the police services on security management and the tendency to privilege experience over qualifications. The fact is that there are limited learning routes into security management and their completion is not necessary to achieve a security role. Worse still, an individual could complete some of the 'top' security courses (academic and vocational) without improving in any way their chances of securing a position in security management. Training and educational achievement are perceived as being of marginal importance in the broader security occupations. Indeed I once attended a careers event for my security students who were bluntly told by a realistic recruitment consultant specialising in security that unless they were backed up by past experience, their qualifications would prove virtually useless in securing them a job in security management.

A study by Hearnden (1993) supports these findings, although the situation has probably improved since, with the expansion of higher education courses in security management. Hearnden found only 8.5 per cent of those surveyed were graduates, as opposed to a national UK average of 13 per cent. For 'A' levels the disparity is similar, with only 24 per cent of managers holding at least one certificate as against 51 per cent of the UK workforce as a whole. Hearnden explains this lack of formal education by the large number of second career entrants from the police and military. Some of the most important vocational and academic courses for security managers which emerged during the 1990s and which will have changed the Hearnden picture are listed in Table 5.3.

A striking finding from Table 5.3 is the low number of higher education courses. There are three undergraduate and five postgraduate

Table 5.3 The main security management qualifications in the British private security industry

National qualifications framework	Framework for higher education qualifications	NVQs	HE examples	Non-HE examples
Level 8 BTEC advanced professional diplomas, certificates and awards	D (Doctoral) Doctorates	Level 5		
Level 7 BTEC advanced professional diplomas, certificates and awards	M (Masters) Masters degrees, postgraduate certificates and diplomas	Level 5	Cranfield University: MSc Security Sector Management Middlesex University: MSc Work Based Learning Studies (Corporate Security Management) University of Leicester: MSc in Security and Risk Management (dl) University of Loughborough: MSc in Security Management (dl) University of Southampton: MSc Corporate Risk and Security Management	
Level 6 BTEC professional diplomas, certificates and awards	H (Honours) Bachelors degrees, graduate certificates. (old Level 3)	Level 4	University of Leicester: BA in Security and Risk Management (dl) University of Portsmouth: BSc in Risk and Security Management	

Table 5.3 The main security management qualifications in the British private security industry – *continued*

National qualifications framework	Framework for higher education qualifications	NVQs	HE examples	Non-HE examples
Level 5 BTEC/HND and HNC BTEC professional diplomas, certificates and awards	I (Intermediate) Diplomas of higher education, foundation degrees, higher national diplomas (old Level 2)	Level 4	Bucks New University College: FD in Protective Security Management; and FD Security Systems Management University of Leicester: FD in Security and Risk Management (dl)	Syl: Diploma in Security Management (Perpetuity)
Level 4 BTEC professional diplomas, certificates and awards	C (Certificate) Certificates of higher education (old Level 1)	Level 4		ASIS: Certified Protection Professional NHSCFSMS: Accredited Security Management Specialist Anubis: Certified Security Consultant
Level 3 BTEC national diplomas, certificates and awards, BTEC Diplomas, Certificates and Awards A Levels		Level 3		Syl: Certificate in Security Management (Perpetuity)

Table 5.3 The main security management qualifications in the British private security industry – *continued*

National qualifications framework	Framework for higher education qualifications	NVQs	HE examples	Non-HE examples
Level 2 BTEC first diplomas and certificates, BTEC diplomas, certificates and awards, GCSEs grades A–C		Level 2		
Level 1 BTEC introductory diplomas and certificates, BTEC diplomas, certificates and awards, GCSEs grades D–G		Level 1		
Entrance Level		Entry		
BTEC certificates in life skills, BTEC certificates in skills for working life				

Notes: The ASIS CPP is not mapped against the NQF and it has been located at Level 4, but some might class it as Level 3. There are several Level 2 courses, but these have been excluded from this table because of the focus upon management level awards. dl: distance learning.

courses. It is also notable that there is no campus-based undergraduate degree open to school-leavers who may wish to pursue a career in the industry (so-called UCAS courses). This lack is accentuated when a comparison is made to computer security. A UCAS search reveals 33 courses in computer security or equivalent at 16 institutions. A Google search of postgraduate courses in computer security or equivalent revealed at least a dozen institutions offering masters level courses. It is difficult to gauge the number of computer security specialists. The British Computer Society Security Group (n.d.) has 300 members. There are probably many more computer security specialists. However, even if there were four times the number of security managers the gap between the two groups of security specialists would still be significant. Even more so given that computer security only emerged as a discipline in the 1980s, while general security managers have a much longer history! The same disparity applies with health and safety management. The Institute of Occupational Safety and Health (IOSH) website lists 12 universities providing both undergraduate and postgraduate courses.

The structure in Table 5.3 also illustrates the absence of qualifications that should provide a licence to practice. Ideally there should be a Level 3 award that can be achieved through a vocational course or higher education award and, where appropriate, with experiential learning opportunities, to provide a gateway to the profession. The old IISec courses (now SyI) could achieve this, but unfortunately because of the attitudes of those in the industry they don't. The NHSCFSMS Accredited Security Management Specialist course is a good example of how such a structure could work. In the NHS to become a security manager (or specialist as they call it) it is necessary to qualify on this course, which takes about 5 weeks and is accredited by the University of Portsmouth as worth 40 Level 1 credits against their BSc (Hons) in Risk and Security Management. This course then has a health care security pathway, which specialists are encouraged, though not mandated, to undertake. Some of these ideas will be developed further in Chapter 9.

The lack of a standard training package that provides a licence to practice security management in the UK has actually been exacerbated by the introduction of statutory regulation through the PSIA 2001. No mandatory training standards for security managers have been established. One could argue that as most security managers are in-house and are therefore exempt from licensing this is not an issue. However, while in the door supervision sector there is licensing of both in-house and contract staff, there are no specific competency standards for the managers of door supervisors. Managers should lead by example. They are major agents for change in the culture and standards of security. Ignoring standards for managers sends out a message of the lack of importance of training and education to a sector where such views are already prevalent. It is interesting to compare the introduction of regulation in the private security industry in the UK with the regulation of health care security in the NHS where a conscious decision was taken at the outset to introduce standards for managers in all NHS trusts, as described above. As we will see below, standards for security managers are mandated in other countries and are of vital importance to the raising of standards overall.

4. Standards and a code of ethics

The final important trait of a profession is the institution of standards and codes of ethics that govern the practice of the profession, that are adhered to and that are actively enforced. Most of the professional associations listed above have standards and/or codes of ethics. However, what is more important is their acceptance and the existence of enforcement structures. It is in enforcement that most organisations fall down, as they have neither the teeth nor the capacity for the role. The fact is that not being a member of these associations doesn't matter. Thus the governance of security managers through codes of conduct and ethics is minimal – if not non-existent – in the UK. Regulation could have a great deal to offer here as well, by setting a statutory framework to enhance standards of ethics and conduct. Unfortunately this has not been on the agenda of the SIA so far.

Security inequity

The final aspect of the weak foundations on which security is built is security inequity. This is an issue that many security practitioners might see as incongruous in the context of enhancing security. It is rarely discussed at practitioner level, although in more academic circles it has become a much debated issue (see Johnston and Shearing, 2003; Wood and Shearing 2007; Loader and Walker, 2007; Reiner, 2007; Kempa and Singh, 2008). Nevertheless in my view, tackling security inequity is essential, not just in terms of furthering social justice, but also in terms of enhancing security in society as a whole (Commission on Human Security, 2003). Below we will consider some evidence of security inequity before articulating why such inequity undermines security overall and needs to be addressed.

It has been well documented that some spatial developments pose challenges to the broader security of society. At one level there has been a growth of what have been called 'gated communities'. These are housing developments on private land where access to the public is restricted. Such developments have expanded massively in countries such as the USA, South Africa and Brazil (Davis, 1990; Blakely and Snyder, 1997; Singh, 2005; Huggins, 2000). However, even in the UK over a thousand such schemes have been identified, some of them guarded by security officers over and above the physical and electronic security measures that are part of the infrastructures of such developments (Atkinson and Flint, 2004). Such communities are largely driven by a desire for exclusivity and for security (Atkinson et al., n.d.).

Gated communities represent the model of opting out from security and social cooperation. As one of the police officers interviewed in a study of gated communities by Atkinson and Flint (2004: 17) noted, 'Overall, there has been a negative impact. It is very much a separate community whose residents use their own [private] schools and shops ... There is very little interaction with other local residents, and it has not brought social or economic benefits.' Nodes neighbouring such gated communities often have minimal levels of security supplied by the state, supplemented in some cases by deviant or vigilante initiatives (Sharp and Wilson, 2000). Through displacement and their own 'thin' security structures they are often at increased risk of crime.

Related in some respects to the development of gated communities has been the expansion of purpose-built shopping and leisure facilities based on private space (Kempa et al., 2004). In some towns and cities private shopping malls have replaced public high streets and such leisure facilities have become the main focus of entertainment. As such places are built on private space the landowner (and his/her agents) can make use of an impressive range of rights. The most important of these include determining who has access and the conditions on which they are allowed to remain on site. Therefore access to shops, outlets, post-offices and so on – frequently providing essential goods and services – has become subject to the whim of the landowner. Concerns have been expressed that the private security officers who enforce order in such locations can use their powers arbitrarily to exclude what are perceived as less desirable groups, such as down-and-outs, the unemployed, teenagers and certain ethnic minorities – creating Charles Reich's feared 'internal exile' (cited in Gray, 1994: 175). Indeed the power of the owners of such private space was illustrated by an incident in a shopping centre in Fareham in the UK in December 2007, where two shoppers who took photographs of relatives outside a Card Factory Shop were asked to leave by a security guard because taking pictures created a 'risk of terrorism' (*The News*, 2008).

Such space has become the subject of legal debate and led to the argument that it should legally be treated as 'quasi-public' or 'hybrid' space, where for example, access should be based upon a reasonable rather than an arbitrary basis (Gray and Gray, 1999a). Nevertheless the case of *CIN Properties v Rawlings* [1995] 2 EGLR 130, illustrated that this is not yet so in the UK. In this case CIN, the leaseholder of a shopping centre, sought indefinitely to ban a group of unemployed youths from the precincts of the centre after unsubstantiated allegations of misbehaviour. The ban was subsequently reinforced with a court injunction. CIN argued that as they were the owners of the property they did not have to show any good cause for denying entry. The Court of Appeal upheld this decision and the European Commission of Human Rights was unable to intervene because the UK – at the time – had not ratified the guarantee of liberty of movement. This decision has effectively given property owners unprecedented power to regulate citizens' freedom of movement, assembly, association and speech. However, the implementation of the Human Rights Act 1998, Gray and Gray (1999b: 50) argue,

> ... will result in a significant curtailment of the estate owner's right in respect of quasi-public property. The incorporation of Convention freedoms will effectively impose on landowners a duty to demonstrate that any exclusion from privately owned areas of quasi-public space is justified on reasonable grounds which do not contravene the guaranteed liberties of the citizen.

Whether such space does legally emerge in the UK remains open to legal debate. Whatever the result, the expansion of such space creates further challenges for those responsible for community safety. Increasing amounts of shopping and leisure time will continue to be focused on such spaces and according to the landowner's safety/security regime. This might be undertaken in partnership with public agencies, or it might not. And even if partnerships are in place, the security of significant locations for the community will be subject to negotiations between partners rather than to the democratic will of the community. Take, for instance, the decision to ban those who wear hooded tops (so-called hoodies) from the Bluewater Shopping Centre in Kent (BBC News, 2005). Even if we were to accept that those who wear such dress are more likely to misbehave (which is unproven) their exclusion will merely divert them elsewhere, where the much less well resourced public infrastructure will have to deal with them. And could this be the thin end of the wedge and open up the banning of other forms of dress, or other groups, actions which could have more sinister undertones?

Many security strategies are forms of situational crime prevention and such strategies are bound to lead to some displacement, depending upon the type of crime, although the extent of such displacement is contested (Reppetto, 1976; Clarke, 2005). If some nodes are better protected than others as a result of possession of greater resources, knowledge, networks and so on their success is likely to come at the expense of another node (in varying degrees), which will see increased targeting by malefactors. In effect, the more efficient nodes are creating 'security pollution' that may afflict their less well protected neighbours. If this were real pollution such behaviour would not be tolerated. Indeed, in most countries, strict regulation prohibits dumping of pollutants on neighbouring lands and polluters are obliged to deal with their own pollutants. Whereas such specific regulation would be neither practical nor justified in the case of nodes where security is effective and malefactors are displaced, such good nodes can nonetheless help the less protected nodes, which is something that will be explored in Chapter 9.

This leads us to consider the implications of allowing 'failing or deviant nodes' to continue. If, the underlying causes of insecurity are not addressed and malefactors are not actively targeted across all nodes, there will be the opportunity for the development of havens of insecurity in which malefactors will be able to base themselves, to 'breed', to develop their tactics and to continue their deviant activities. These will eventually return to the better protected nodes, prompting further advancement and expenditure in a security arms race. Research in South Africa for the World Bank has illustrated how wealthier nodes are targeted for burglary by neighbouring deprived nodes (Demonbynes and Ozler, 2002). Such arguments have also been made vis-à-vis terrorism and 'failing' states, which have been classed as havens for terrorists, enabling them to perpetrate their attacks elsewhere (Newman, 2007). Ultimately, it is in the interest of the better protected nodes to ensure all nodes are protected. As Clarke (2005: 58) has argued, 'In any case, wealthy people have a strong interest in an orderly society, since order is a basic requirement for the production of goods and services that they consume and the generation of wealth from which they profit.'

Next there is a plain social justice argument against security inequity. As discussed above, security is a basic human need and minimum levels of security should be a fundamental aim of the organisations of governance (Commission on Human Security, 2003). Every person, no matter where they live or go is entitled to minimum levels of security and it is fundamentally wrong that there should be such deep-seated differences within society, particularly when the better security in one node may actually worsen the security of those in another.

It is important to note that it is not possible for a person to spend all his/her time in well protected nodes. People travel across nodes all the time, as do the goods and services of corporate nodes. It is therefore difficult to avoid at some point either visiting a less protected node or coming into contact with a malefactor based in less protected node. It is therefore in the interest of the better protected nodes to ensure that their people, assets and so on are protected as well when they travel beyond their node as they are within it.

Finally, a growing body of research on 'signal crimes' has illustrated the way in which incidents of insecurity can send out signals beyond the node where they take place, leading to disproportionate perceptions of actual risk and impacts on behaviour (Innes and Fielding, 2002). The 2006–07 British Crime Survey (BCS) confirmed the statistics in the previous survey and indicated that the chances of becoming a victim of crime are the lowest since the BCS began. Since the peak in 1995 there were over 8 million fewer crimes and the risk of victimisation had gone from 40 per cent to 25 per cent (Nicholas et al., 2007). Yet the same research found that nearly two-thirds of those surveyed continue to think crime is rising. A brutal murder, rape or abduction affects people's perceptions of security, including those who live and move around in areas where such incidents are least likely to occur. Thus failure to address security in some nodes ultimately impacts upon the better protected through greater feelings of insecurity. If the better protected are to sleep more soundly they need the security deficits in the less protected nodes to be addressed.

Tackling the inequity of security between nodes across society is therefore a fundamental part of a macro level strategy to enhance security. The ultimate foundations of security will be unsound if there are many nodes with significant security problems. There are a wide variety of measures that can be pursued to achieve this, from the sharing of knowledge, skills and resources and, most radically, through some redistribution. These ideas will be developed in more depth in Chapter 9.

Conclusion

This chapter explored the weak foundations for the development of effective security systems in society. It has shown that regulation for those delivering security in the UK and in many other countries is ineffective. Systems range from the merely inadequate to those that actually inhibit effective security. The chapter then considered why security management needs to be a profession and the weakness of professional characteristics within the sector. The orientation of security managers, their professional associations, their qualifications and the codes of governance were all shown to be inadequate. Finally we looked at the more radical concept of security inequity and showed how failing to address security across the whole of society ultimately undermines security in the better protected nodes, creating a problem that needs to be tackled. Part II has tackled the weaknesses in doing security. Part III will examine how security should be done. It will start by setting out a nodal-level holistic security model before going on to examine how effective security can be created at a nodal level and ultimately how the foundations of security can be rebuilt.

Part III
Doing Security

6
Creating a Model Holistic Security System

Introduction

This part of the book is concerned with the process of identifying how security should be done. Below, this process begins with a model for the creation of a holistic security system at a nodal level, whereby the problem is defined, a solution is tailored to address it and a process established to enable ongoing evaluation to tweak the solution. The model proposed draws upon the principles of modern risk management and good practice in policy formulation, but it is further enhanced by the incorporation of some of the latest innovative thinking in security. The creation of the model begins with a discussion of the strategies and challenges of defining the problem in a node. The chapter then explores how a solution can be developed to address a particular problem, utilising both risk management strategies and economic decision-making tools to influence the system. The chapter then looks at some initiatives that can be used to ensure that the strategy remains current and is still applicable to the problems the node faces. The final section looks at a case study of the National Health Service Counter Fraud and Security Management Service (NHSCFSMS) to illustrate best practice in developing a holistic security system. Before we embark upon this, however, it is important to outline the ways in which security can be the new tool for competitive advantage for organisations.

Security as competitive advantage

Briggs and Edwards (2006) have argued that security could become the new source of 'competitive advantage'. In the related area of fraud, Jim

Gee, from KPMG Forensic and former Head of the NHSCFSMS has also argued that an effective counter-fraud strategy could provide competitive advantage to an organisation. The thinking behind these views goes as follows. Many organisations, particularly in the commercial sector, have introduced waves of new business strategies, from outsourcing to flat management structures and so on (Porter, 2004). However, improving security capacity – which can have an overall benefit on the bottom line by reducing losses – has been neglected by many organisations. Perhaps testament to this are the very few MBA courses that offer courses on security and counter-fraud (Gill et al., 2007). Given that the scale of losses from fraud and insecurity can account for significant sums in financial terms, more effective security strategies have the potential to bring substantial return on investment, possibly more so than many other more conventional business strategies. This is not just a potential benefit to the commercial sector in reducing costs and increasing profits, but also applies to many public sector organisations, where the reduction in losses can be freed up to provide more resources for better public services. The potential benefits can be demonstrated with reference to the NHSCFSMS and its holistic approach to tackling fraud. It reckons to have saved the NHS £811 million from fraud, which amounted to a 12 to 1 return on the investment in the NHSCFSMS (NHSCFSMS, 2007).

A holistic security model

Developing a holistic security model demands the creation of a system based on the following stages: defining the problem, developing a system focused upon the problem, and creating mechanisms to ensure that the nature of the problem is monitored as is the effectiveness of the strategies targeting it. This model is set out in Figure 6.1, and some of the more specific tactics and strategies are outlined below. These will later be explored in more depth.

The model has been influenced significantly by the NHSCFSMS model for tackling fraud and security, which will be discussed later. It also mirrors most of the Secured by Design 'Secured Environments' seal of approval issued by the police (in partnership with the Perpetuity Group) to organisations that meet its six principles of protecting themselves against crime. These principles are:

- A commitment to create a secure environment
- Understand the problem

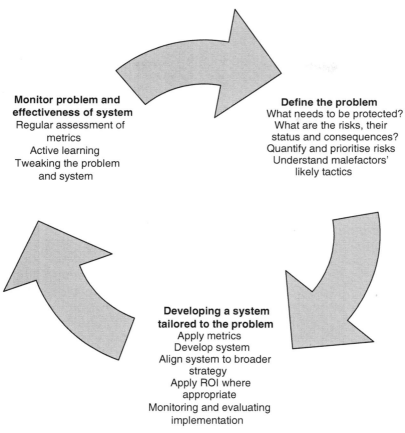

Figure 6.1   A model for the development of a holistic security system

- Develop a response
- Manage the response
- Implement the response solution
- Evaluate the response

(Secured Environments, 2007)

Defining the problem

The first task before any risk analysis or strategies can be developed is to identify within the node what needs to be protected and what the priorities should be. These can be categorised as people, property, information and reputation. The extent to which these four areas need to be protected vary from node to node. For example, reputation will be of greater significance to a security manager working for a company with a highly desirable brand than to an individual working for a local authority. Once the areas that need protecting have been identified the next part of the system is risk identification. The basic principles of this will shortly be explored, but before this is done it is worth briefly defining risk itself.

There is much debate over what constitutes risk and there is not enough space here to engage with it (those interested are referred to Adams, 1995 and Borodzicz, 2005). For the purposes of this book the Royal Society Study Group (1992: 2) definition of risk will be used, which is, 'The probability that a particular adverse event occurs during a stated period of time, or results from a particular challenge.' A risk therefore has three components: a future event or a hazard; the probability that the hazard or future event will occur; and the adverse consequences should the hazard or future event occur.

In the context of nodal security a risk is any adverse event with a probability of occurring against the node that, should it occur, will have some kind of negative consequence. In Chapter 1 some of the common risks against which nodes protect themselves were illustrated: crimes, protests, deviant acts, accidents and incursions by non-denizens. In varying degrees these form the risks that different nodes face. However, the very different shape of these risk profiles acts as a significant influ-

Table 6.1   Risk profiles in four different nodal contexts

A supermarket	An airline	A hospital	A housing estate
Shop-theft	Hijacking	Violence against	Anti-social
Staff theft	Bomb	staff and patients	behaviour
Burglary	Missile attack	Theft of drugs and	Burglary
Criminal damage	Violence against	drug abuse by staff	Criminal damage
Arson	staff and	General theft	Car crime
Robbery	passengers	Fraud	Assault and
Fraud	Drug abuse by staff	Child abductions	violence
Violence against	Counterfeit parts	Loss of information	
staff and	Intruders	Theft of hazardous	
shoppers	'Ordinary' risks of	substances	
Blackmail	theft, burglary etc.	Car-park crime	
Terrorism		Stalking/voyeurism	
		Criminal damage	
		Bomb-threats	

ence upon nodes, in turn shaping (amongst other factors) the security system. Table 6.1 shows the common risks faced in four different types of node.

It is also important to denote different risk environments according to the nodal context. Figure 6.2 below shows two continuums. The horizontal line covers the level of security required, ranging from minimal to 'absolute security'. The more severe the potential consequences of adverse events occurring the greater the level of security required. Thus the protection of politicians and the buildings they work in (that is, the political infrastructure) requires absolute security, and a local museum, with an average range of artefacts, or an ordinary housing estate requires minimal security. The vertical continuum ranges from environments where there are strong commercial pressures, such as supermarkets, to those where there are no such pressures, for example, military bases and the political infrastructure. These continuums cross and there are some nodal contexts that require absolute security, but which also operate in highly competitive markets. Aviation security is an example of this, where the risk posed by terrorists requires absolute security, but where airlines and airports also operate in highly competitive commercial environments in which costs are very important (Hainmüller and Lemnitzer, 2003; National Commission on the Terrorist Attacks upon the United States, 2004). By contrast in retail environments the level of security is often minimal, because most of the consequences of potential security failure are financial losses (although violence against staff is a major

Figure 6.2 Four models of security

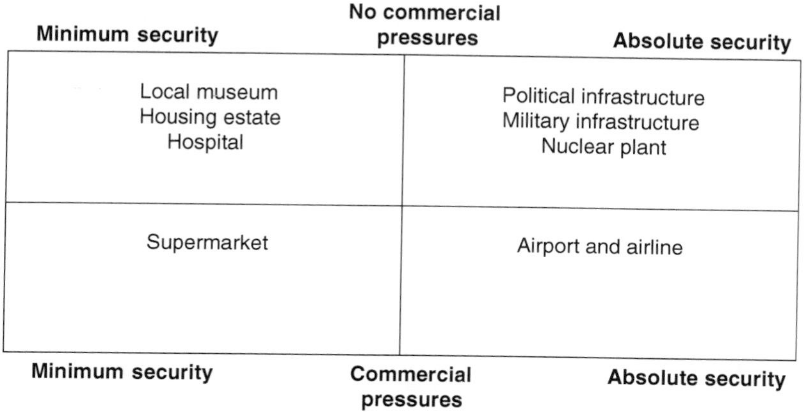

Minimum security	No commercial pressures	Absolute security
Local museum Housing estate Hospital		Political infrastructure Military infrastructure Nuclear plant
Supermarket		Airport and airline
Minimum security	Commercial pressures	Absolute security

concern in some retail outlets) (Beck and Willis, 1995). At the other end there are nodal contexts where the commercial pressures are minimal or non-existent and where levels of security differ. In general, hospitals do not require absolute security and commercial pressures are less strong (Colling, 2001), whereas the US White House requires 'absolute security' although there are no commercial pressures (although lesser pressures of public access to politicians are an influence) (White House Security Review, 1995).

These different environments have important implications for the security system that emerges. Commercial pressures inevitably lead to greater focus upon costs and some security managers need to apply security within a much more constrained environment. In contexts where absolute security is required this can lead to conflicts. For example Hainmüller and Lemnitzer (2003) have argued that the airline lobby and the desire to eliminate 'bureaucratic' security costs in America resulted in inadequate security systems that compromised the ability to deal with the terrorist threat. Commercial pressures in retail security by contrast have few implications beyond the bottom line. At the other end of the spectrum absolute security in a non-competitive environment can result in an oppressive security system, such that, for example, it becomes difficult to access politicians, that is, to visit the political infrastructure. The opposite can also occur in nodal contexts where minimal security is required with limited commercial pressures, and security can be neglected, as happened in some areas of risk within the NHS, such as violence against staff, before the formation of the NHSCFSMS (National Audit Office, 2003).

In assessing risk it is also vitally important to understand the malefactors who are likely to perpetrate risk events. As Chapter 3 illustrated there is much to be gained from an understanding of how different types of malefactors go about their work. This information can be used in the development of the holistic model, in pursuing the solutions to counter the identified risks.

Once the risks and the nature of the malefactors have been defined, it is possible to explore the management of the risk process. This can be divided into four distinct stages and the level of sophistication of these can vary significantly between nodes. The first stage involves the identification of risks faced by a node and a number of different strategies exist to enable this to take place. Second is the risk estimation stage where the probability of identified risks occurring is established. In the third stage risk evaluation takes place. In this stage, the tolerability of risks is assessed vis-à-vis their likelihood of occurring and the potential consequences if they do occur. Each of these stages will be discussed below.

Risk identification

The use of systems to manage risks has permeated the corporate sector and trickled down to many parts of the public sector (Johnston and Shearing, 2003). A wide range of risks of which security is just one are identified and managed by these systems. Central to the mentality underlying risk management is a future-oriented approach that seeks ideally to prevent or at least to reduce the likelihood of particular events or behaviours from occurring. Such ways of thinking have become very attractive to some writers on security who compare such approaches to the predominantly reactive and past-oriented strategies used by many public sector security agencies – though this tendency is weakening (Johnston and Shearing, 2003). Risk assessment deals with the identification of risks and their consequences should they arise. There are a number of techniques in existence for the identification of risks. Some of the most popular, identified by Frosdick (2007), are listed in Box 6.1.

One of the fundamental problems in attempting to identify risks is that methods are often based on hindsight and past incidents. Such approaches have a number of weaknesses. First, organisations frequently fail to gain any understanding from passive learning from previous non-disastrous events. For instance it was well known in NASA that there was a problem with the 'O' rings that sealed the solid fuel rocket boosters on the Challenger space shuttle, yet no active learning had led to this risk being addressed (Toft and Reynolds, 1997). Second, risks are often identified on the limited basis of the organisation's own experience, or the experience of similar bodies within the same industry. Toft and Reynolds (1997), however, argue that many incidents display similar characteristics and therefore learning could be pursued from a much wider range of organisational experience. Finally hindsight cannot predict new risks. The events of 11 September 2001 in New York illustrate this. Few or no organisations would have predicted the risk of an airliner being deliberately flown into a building by suicidal terrorists.

Another problem for risk identification is cultural bias. Cultural theorists Thompson and Wildavsky (1986) have shown that different organisations accept and reject information in different ways. As a consequence some risks are neglected. For instance, few risk assessments of Boeing 737 aircraft would have thrown up the possibility of a pilot

Box 6.1   Risk identification techniques

Group creativity techniques (often known as brainstorming): A group of people sit round a table and are encouraged to think of any risk – however far-fetched.

Inspections: An approach often used in health and safety is inspection where a person simply surveys an area for potential hazards using a pre-produced questionnaire.

Checklists: In some organisations checklists are provided covering common areas where there might be risks such as windows, doors, where valuables are kept etc. and the area is surveyed according to these lists.

Failure modes and effects criticality analysis (FMECA): Under this approach the system been reviewed is converted into a diagram based upon the distinct parts of the process. The diagram produces the broader system and the underlying sub-system. Failures are then considered with the implications for each sub-system.

Fault tree and event tree analysis: Under fault tree analysis you start with the event at the top of the tree and work down. So with theft of drugs from a secure cabinet you would work down, identifying what might cause the theft to occur. There would need to be a criminally motivated offender and an opportunity for them to access the cabinet, such as an open door or ineffective lock. The tree continues until it is no longer possible to undertake any further analysis. With event tree analysis the process is reversed. So using the theft again the questions might be: is the cabinet open? Are the locks working? Is there an alarm and does it work? And so on. The answers to these lead to a variety of consequences.

Official investigations: Frequently disasters and accidents result in investigations which result in recommendations that seek to prevent their recurrence. These often clarify risks and identify actions to manage them.

Source: Adapted from Frosdick (2007).

shutting down a wrong engine and not being advised of his error by fellow crew members because of the social and hierarchical cultures in place. Yet in the Kegworth air-crash in England that is exactly what happened (Weir, 1996: 118). Ultimately those responsible for the identification of risks will have their own cultural biases which may lead to a failure to identify some risks or to risk-priorities being wrongly estimated.

Risk evaluation

Once the risks have been identified the next stage is to estimate the probability of them occurring, which in many cases involves detailed statistical analysis. Quantitative methods are often used to predict the probability and this approach is usually termed quantified risk assessment (QRA). Some organisations use very basic criteria and merely rate probabilities as low, medium or high. Linked to the probability of occurrence is the need to evaluate the potential consequences. Again some organisations pursue quite sophisticated analysis while others simply rate the consequences as low, medium or high. This process can lead to the realisation that some risks cannot be wholly eliminated, only reduced. This in turn leads to measuring the extent of certain risks.

A good example of such detailed risk analysis has been the NHSCFMS risk assessments of different types of fraud, which have been conducted on a regular basis. These have covered, among others, prescription fraud, dental contractor fraud and optical patient fraud. To illustrate how they work the risk management project for prescription fraud will be assessed. In the NHS in the UK most people have to pay a flat-rate fee of £6.85 for a prescription. There are, however, numerous groups entitled to free prescriptions, among them the unemployed, pregnant women and people with certain categories of illness. Some people fraudulently claim free prescriptions. To gauge the extent and risk of this fraud a statistically valid sample – reflecting a cross-section of ages, class, race and so on – of 3200 exemption claims was analysed. These claims were classed as correctly exempt, fraudulent or incorrect (no proof either way). From this the national picture was extrapolated using statistical analysis (Smith, 2006). With some slight changes in methodology these exercises were continued. When conducted on a regular basis using the same methodology they also provide a good means of assessing how effective the strategies put in place to reduce the risk were. Indeed on prescription fraud these risk management exercises were able to show a saving of £117 million between 1998–99 and £47 million between 2003–04 (NHSCFSMS, 2005).

The NHSCFSMS is typical of a number of organisations that are investing in 'metrics' (detailed statistical information about the organisation) to enable more accurate decision-making (Gill et al., 2007).

The development of such systems to produce data is being used not just to evaluate risks, but also to measure the value of the security strategies that are applied to address them. For some security managers this provides them with much more effective 'tools' to talk the language of the board, demonstrate value and achieve the outcomes they desire for their security departments.

When considering risk evaluation it is important to note the phenomenon of risk homeostasis or risk compensation theory. For Wilde (1982), decision-makers will subconsciously balance risks against potential gains to determine acceptable behaviour according to their own personality, economic circumstances, culture and so on. Adams (1995), who has taken Wilde's ideas further, argues that we all have an internal risk 'thermostat' setting our preferred level of risk. Some people tolerate higher risks than others. Thus a pensioner might perceive of driving at 100 mph as too risky for a pensioner, while a middle aged salesman who drives extensively might find it quite acceptable. Linked to this is that measures to reduce risk may actually lead to increased risk-taking. Thus the introduction of the compulsory wearing of seatbelts could lead to an individual driving faster than before because wearing a seatbelt makes them feel safer.

In many fields of risk management the data on unwanted events are either relatively recent or unrepresentative or unreliable. While there are some areas in which extensive records enable more accurate predictions, such as the weather (where records go back many years and enable predictions to be made of the risk of flooding, hurricanes, cyclones and so on – although global warming is undermining the usefulness of past data to some degree), in many areas of social life, the history of data is much less reliable. For instance with the emergence of new technologies, such as the internet, over the last two decades new risks have emerged, such as computer 'viruses', 'worms' and hackers. However, because of their relatively short history and the swift pace of technology it is difficult to use past data to estimate the probability of such incidents occurring. Also in many organisations and spheres of life the recording of incidents is not always accurate. Some incidents might be recorded periodically or rarely. Crime, for example, for various reasons, is not always reported or recorded. Thus the apparent risk of burglary in a particular location might not reflect the reality of the risk because the calculation is based upon inaccurate recorded figures and not the actual number of incidents.

It is also dangerous to presume, particularly with quantified risk assessment, that there is some kind of scientific objectivity to the process. As some of the above examples have illustrated not all the risks might be identified, the estimation of their occurrence might be flawed and the management process may also be inadequate. In many circumstances two teams could be asked to undertake a risk assessment of the same situation and would come up with two sets of results.

This is where Adams' (1995) distinction between three kinds of risks helps us in interpreting situations. There are risks perceived through science, where there is general agreement. For example, the risk of a bridge collapsing in specific stressful situations or the chances of dying from exposure to radiation. These types of risk have the greatest degree of scientific objectivity. Then there are directly perceptible risks, such as driving a car at 100 mph, climbing mountains and so on. Attempts to manage these risks are thwarted by individuals insisting on their own rights to make judgements. Finally there are risks where scientists do not know or are unable to agree on the risk. Global warming, for example, where there are competing views. With the latter two types of risk, subjectivity dominates over objectivity.

Defining the solution

Once the extent of the problem has been defined and the nature has been determined, it is necessary to define a solution. This is also founded in the principles of risk management.

The management of risk

Risk management strategies are strongly influenced by the nodal context as outlined in Table 6.1 above. In some nodes, such as the protection of the US president, such is the importance of protecting an asset that the costs are not a significant issue. In many nodes, however, the process of deciding what strategies to employ to manage risk is bounded by an economic framework; in some cases, the potential costs of the risk might not warrant the expense of actually trying to deal with it. A tool that can be applied in reaching these decisions is return on investment (ROI). This is generally seen as net income divided by investment, whereby the likely increases in net income from the cost of the investment can be calculated (Garcia, 2006). In the security context it is more the reduction in losses (therefore increased net income) set against the costs of the investment. A simple example to illustrate this could be that of a security manager in a department store which is experiencing increased shoplifting. The security manager might think that the introduction of an electronic article surveillance system (EAS) would reduce the level of shoplifting, but for the investment to be worthwhile, the reduction in the overall losses from shoplifting must amount to more than the costs of introducing EAS. The central question will be, is there an ROI? If yes proceed, if no think of another strategy. Gill et al. (2007) also advocate the importance of aligning the security system with the overall strategy of the organisation, which also contributes towards making a stronger case for security.

There are a number of problems to be aware with in this approach (Beck, 2007). As was illustrated above, there are inherent problems in calculating risks and there are also challenges in recording them. For example, Smith (2006) cites research by PriceWaterhouseCoopers which demonstrated a sound ROI for the introduction of EAS. However, he notes that this was based on accepting inventory shortage statistics as an accurate measure of shop-theft, which in many cases would not reflect factors such as stock control procedures or other means by which stock could have disappeared: damaged, out of date or stolen by staff (where EAS would have no impact).

Some security departments also have great difficulty making the case for ROI. As was discussed in Chapter 4, security managers can find 'talking the language of business' difficult. This can become an even greater challenge because of inherent difficulties in making a case using ROI for many security strategies. For example, if a security department is working effectively and there is apparently 'no problem', requests for investment are often ignored or underplayed. Smith (2006) cites the example of an access control system. It would be very difficult to assess how many potential incidents have been deterred and what costs might have been caused had intruders breached security. If the system is due an upgrade, how can a security manager make a valid case using ROI without resorting to fantasy figures? There is also another side to the ROI decision: if new measures are introduced, might malefactors be displaced to commit different crimes that might prove more expensive to the organisation? Perhaps a more appropriate way to apply ROI to security is not to use it in support of particular measures, but to measure it against the totality of the department and to benchmark against comparable departments or organisations. Although such data sharing and investment in such research are rare in the security world and rarely publicised.

Looking at ROI, security managers also need to be aware of the latest research on the effectiveness of different types of security services and products. There is a growing body of literature which gives useful insights. For a few examples on CCTV see Gill and Hemming (2004) and Gill and Spriggs (2005), on electronic article surveillance see Bamfield (1994) and Handford (1994), on security guards see Button (2007a), on radio frequency identification, Beck (2006), and on explosives detection, Baldeschwieler (1993) and Oxley (1993). *Security Journal* also regularly publishes articles of this type.

Some organisations have quite sophisticated decision-making models. For example the Post Office has a model to protect its post offices which is based on a complex range of data, such as proximity to a motorway, the social demographics of the area, past incidents and so on, which is scored and a range of measures brought into play according to the result. Many other large organisations have sophisticated risk models such as this. The principal strategies to manage risk are listed in Box 6.2. The most common strategies relate to risk reduction. A much more detailed analysis of the variety and development of such strategies, not to mention ensuring their maximum impact, will be the subject of chapters 7 and 8.

Once the system has been decided upon the next stage is implementation. Many security systems fail because of poor implementation. The problem might be a poor project manager, stakeholders might not be properly consulted and integrated in the implementation and senior managers might lack ownership and therefore commitment to the system (Secured Environments, 2007). It is therefore vital that any new system is implemented using appropriate project management principles, that there is effective consultation with stakeholders and ownership at a senior level.

Monitoring the nature of the problem and the effectiveness of the system

Once the system calculated to address the problem is up and running it becomes important, first, to continue to assess the problem and to amend the system accordingly, and second, to monitor its effectiveness. This means regularly repeating stage one above. Central to ensuring that this is done in the most effective way is the use of data monitoring systems to produce metrics which enable constant assessment and more detailed periodic analysis. Gill et al. (2007) advocate the use of metrics by security managers and provide some examples (see Table 6.2).

Another very important strategy that needs to be pursued in order to maximise the effectiveness of the system is to conduct learning. In the

Box 6.2   Principal risk management strategies

No action: It might be decided that the risks of an event occurring are so unlikely and inconceivable that no action is needed to be taken to deal with the risk. For instance post-11 September some owners of prominent high-rise buildings may have considered a new risk to be suicidal terrorists in a hijacked aircraft flying into them. However, while now a recognised risk many may have decided that as it is so unlikely that no action will be taken to strengthen the building.

Risk avoidance: The risk of a particular course of action may become so high that it is decided to avoid the risk altogether. For instance the growing perception of the risk of terrorism post-11 September has led to many business organisations refusing to send staff to countries such as Pakistan or withdrawing those who were already there. Similarly organisations with properties in some countries at risk of terrorism may have decided to relocate to safer locations (George et al., 2003).

Risk reduction: One of the most common measures is to introduce strategies to reduce risks to a tolerable level. This is the purpose of security systems, and some of the tactics that can be used to achieve this will be explored in chapters 7 and 8.

Risk transfer: Even when risk reduction measures have been introduced there is invariably still going to be a residual level of risk. To cover this an option might be to transfer the risk to someone else. For instance a hospital with a car park where cash needs to be collected by its staff who have been the victim of a number of robberies may decide to transfer the risk to a security company by paying them to collect the money.

Insurance: One of the most common strategies is to simply insure against the risk happening.

Source: Adapted from Frosdick (2007).

exploration of the causes of security failure in Chapter 2 it was demonstrated that in many cases there were precedents from within the organisation or from other bodies which could have given insights to decision-makers and helped them to deal more effectively with a

Table 6.2   Example metrics advocated by Gill et al. (2007)

Aspect of security	Suggested key metrics
Losses during supply chain transportation	Value lost in £ per journey. Value lost in £ as a percentage per cost of goods transported.
Theft by staff	Number of successful searches as percentage of searches conducted in each calendar month. Value £ of property recovered.
Reduction of fraud by suppliers	Number of discrepancies per delivery. Value in £ of discrepant deliveries.

Source: Adapted from Gill et al. (2007: 83).

potential risk. Chapter 2 cited the three different types of learning advocated by Toft and Reynolds (1997): organisational, iconic and isomorphic. The latter, they argue, is the most important because, to quote Bismarck, 'Fools say they learn from experience. I prefer to profit from others' experience.' Toft and Reynolds (1997) break down isomorphic learning into four types, which can also be applied to security. Indeed many security managers who have responsibility for safety, disaster management and the like may already use similar learning strategies.

Event isomorphism

This is where the *hazard is the same* but *the events* that cause them *can be different*. Using the Clapham Junction rail disaster of 1988, Toft and Reynolds (1997) illustrated that *two types of events* can cause a train to enter a section of track that they should not be on. First, a driver could pass a red signal, creating a 'signal passed at danger' (SPAD) incident. Alternatively, the signal could fail to change to red otherwise known as a 'wrong sided signal failure' (WSF). Both these could occur as a consequence of human or technical failure. While both are equally dangerous, SPAD had received much more attention than WSF and the inquiry into the disaster argued that an opportunity had been missed to reduce the risk of such an incident occurring again, because the hazards of WSF had not been fully discussed or addressed. These same principles can be applied to security. When security failures take place it is important to understand what went wrong and learn from that experience. For example the security failures involving different protest groups against the Palace of Westminster, Buckingham Palace and Windsor Castle discussed in Chapter 2 should have alerted the managers of any high profile public building of potential security risks and the many different tactics used to breach security.

Cross-organisational isomorphism

This is where *similar organisations* can learn from one another's experience. Toft and Reynolds (1997) cite the example of a Turkish Airlines DC10 which crashed in Paris in March 1974. It crashed as a consequence of a rear cargo door having been fastened incorrectly. This opened at 12,000 feet as a result of vibration, causing rapid decompression which in turn caused the cabin floor to collapse, cutting all means to control the aircraft. However, this was not the first time this had happened. In 1972 an American Airlines DC10 in Canada experienced the same problem, although on this occasion not all the controls were lost and the crew was able to land safely. Hundreds of reports of problems related to closing the cargo doors were also filed with the manufacturers. Perhaps if Turkish Airlines were more attuned to these reports, the disaster could have been prevented? Similarly arguments can also be applied to security. Thus the example of the Gardner heist in Chapter 2, should have motivated other galleries to investigate their own procedures regarding the admittance of visitors at night, the location of alarms, the quality of training and supervision and so on.

Common mode isomorphism

With this type of learning the organisations belong to different industries but *share common tools, components, materials, techniques, procedures* and so on. An example of this would be the use of polyurethane foam in the furniture and aircraft industries. Disasters resulting from fires in this foam in the furniture industry were heeded by aircraft manufacturers, who chose different materials for use in planes. The 'components' of security systems, such as security guards, CCTV systems, intruder alarms and so on all provide a basis for common mode isomorphism. For example, in November 2007 concerns were raised concerning certain locks on PVC doors which can be 'bumped' open simply by placing a similar key in the lock and tapping it. Anyone with such locks can act on this information and replace them with more effective devices or use a product such as Pickbuster (Info4security, 2007).

Self-isomorphism

This is where the organisation is so large that it has *sub-units* that provide the basis for learning from incidents which take place within the same body. Thus the example of the breach of security at Prince William's birthday party discussed in Chapter 2 should have provided the Metropolitan Police with a huge amount of information that should help ensure that such an event doesn't happen again. It is also important that the isomorphic learning is active and not passive. For instance, there were existing concerns over the location of the security desk at the Gardner Museum but they had not been actively addressed.

The prevention of security failure can be improved by isomorphism at a general and specific level. In the prevention of future security failure it is important to learn from past experience *within and beyond* the organisation. This experience can then be enshrined in the plans to prevent and deal with dangerous incidents.

Isomorphic thinking and security: the case of maritime security

An example I want to use to illustrate the potential benefits of this approach is maritime security and the terrorist threat. Anyone working in the maritime security field should be using isomorphic approaches as a means to enhance security. The past experience of attacks by new terrorists suggests that they will seek to maximise casualties through the explosion of devices in areas where they will secure significant publicity and where they are confident they can beat security measures. Maritime security measures are much weaker than those in aviation (and we have already seen how ineffective these can be). Maritime security offers targets in the forms of cruise ships or ferries with large numbers of passengers. The experience of the Herald of Free Enterprise (187 dead), the Estonia (800+) and Al-Salam Boccaccio (1000+) demonstrate the potential scale of the risk if the main doors are left open or fail. A successful attack on this scale would bring huge publicity. On international roll on roll off ferries there are only limited security measures and on internal ferries there are even less. Building upon the techniques of isomorphic learning measures to enhance the screening of vehicles travelling onto such ferries should be urgently considered. It is hoped that this is being taken seriously by ferry operators and maritime authorities.

Now that the model for a holistic security solution has been outlined the next section will examine the NHSCFMS where one of the best examples of a holistic security system has been implemented.

Case study: the NHS Counter Fraud and Security Management Service

There are many organisations that pursue a holistic approach to security, but there is one that deserves particular mention: the NHSCFSMS

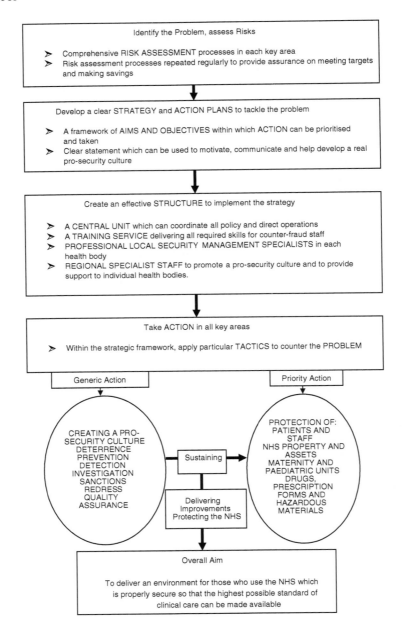

Figure 6.3   The security business model for the NHS

Source: Adapted from Department of Health (2003: 11).

has pursued a holistic and strategic approach to tackling fraud since 1998 and, more recently, has also embraced security (2003) (Department of Health 2003). Security in the NHS has been a major concern because of violence against staff and patients, loss of assets through theft and fraud, as well as rare but high profile incidents where babies have been abducted from maternity units.

Figure 6.3 sets out the security business model, which is almost identical to the counter-fraud business model. It begins with the need to identify the problem through a proper risk assessment. The next stage is to develop a strategic plan to tackle these problems along with a central unit to implement it. In the generic action zone come the strategies to reduce the risks. Central to this is creating a *'pro-security' culture'*. This is reinforced by the other strategies that will be discussed, along with an NHS campaign on security issues. *Deterrence* (or rules and governance) is promoted through new policies on prosecution of offenders in the NHS and the communication of these policies. Thus the NHSCFSMS has pursued successful prosecutions of those who have committed acts of violence where the police and Crown Prosecution Service have refused to continue. The aim is to make clear to potential offenders that criminal prosecution or lesser sanctions are likely if violence is used. The next part of the NHS system is *prevention* (opportunity reduction and design, and technology) where certain areas have been re-designed and there has been the introduction of security strategies such as CCTV and security officers.

Another important part of the system is *detection*, as in the past many security incidents were not reported. Accurate detection is important not only to measure how successful the system has been, but also to enable the targeting of areas of concern because of the importance, as discussed above, of tackling 'spirals of decay' (image and reputation). The NHS system then advocates *investigation* of all incidents followed by *sanctions* and *redress* (image and reputation, rules, governance and sanctions). It is the aim of the NHS to investigate incidents and pursue the most appropriate sanction. This might include all or some of the following: criminal prosecution, civil remedies, termination of employment, removal of licence to practice and the pursuit of financial compensation. Successes in this area are promoted to further contribute to the 'pro-security' culture.

As the security system was only introduced in 2003 it is too early really to evaluate it. However, with fraud, where the system has been in operation since 1998, as noted above, statistics for 1999 to 2006 show a saving of £811 million to the NHS, which amounted to a 12 to 1 return on the investment in the NHSCFSMS (NHSCFSMS, 2007).

Conclusion

This chapter has set out a model for developing a holistic security system. This model can be divided into three parts. The first encompasses strategies to identify the problem such as risk identification, evaluation and so on. The second part involves the development of strategies to address the problem. This involves risk management strategies. Finally it is important to monitor the problem and to adapt the system accordingly as well as to engage in active learning. The chapter ended with an examination of the NHSCFSMS and their holistic security system. This model holds together the framework for more focused action that will be explored in the next chapter using Lukes's three-dimensional conception of power as a means to achieve outcomes. This will be followed in Chapter 8 by an exploration of how the human element in the security system can be enhanced.

7
Making it Never Happen

Introduction

In this chapter we begin to map out the strategies at a nodal level that can be pursued to create a security system that will reduce the risk of crimes and other deviant acts occurring, or even better, make them never happen. The aim will clearly depend upon the nodal context. In a shop the aim will be to reduce shoplifting to an absolute minimum without undermining potential sales, but in an airline the aspiration will be to make hijacking never happen. This chapter will start by setting out a three-dimensional theoretical framework based upon Lukes's conception of power. Ultimately security systems are about power because they are concerned with getting a to do x or not to do x. The chapter will continue to explore some of the three-dimensional strategies (some of which are also two-dimensional). The next chapter will pick up on the most important first and second-dimensional strategies, which are largely rooted in the human element (usually but not always a security officer).

Here the analysis will be divided between strategies aimed at changing malefactors' behaviour (turning malefactors and potential malefactors into good denizens either permanently or temporarily) and strategies aimed at refocusing the behaviour of malefactors (turning malefactors who will not become good denizens towards more benign activities, reducing their activity or deflecting them elsewhere). The chapter will end with a case study of El Al, which, it is argued, provides a good example of an organisation utilising many third and second-dimension strategies. This and the subsequent chapter are also very important in showcasing some of the many examples of research from different disciplines that are of use to the security manager.

Security systems are about power

Security systems are ultimately about getting people to behave or not to behave in a particular way. They are about achieving outcomes and therefore they are about power. The security decision-maker, *A*, wants *B* to do something or not to do something. To achieve an outcome a security manager has a wide array of tools at his/her disposal and a useful way to conceptualise this is to use Lukes's (1974) three-dimensional conception of power.

At the base level power seems a relatively simple concept: the ability of *A* to get *B* to do something they otherwise would not do. This is only part of the picture, however, and this first dimension, as Lukes (1974) would call it, forms the foundations of less visible forms of power. For Lukes there are another two dimensions to power, the second of which involves a critique of the first and is where *A* prevents an issue of conflict from emerging so that *B* pursues a course of action that if that issue had arisen might have been pursued differently. As Lukes (1974: 20; emphasis in the original) writes:

> ... the two dimensional view of power involves a qualified critique of the *behavioural focus* of the first view and it allows for consideration of the ways in which *decisions* are prevented from being taken on potential issues over which there is an observable *conflict* of *interests*, seen as embodied in express policy preferences and sub-political grievances.

Such power was also recognised by Foucault (1977) to inhere in Jeremy Bentham's panopticon, where a prison constructed to enable the surveillance of prisoners without them knowing if they were being watched meant that conformity was their only option. The third dimension provides a further critique of the earlier two views and is a consideration of ways in which actors' decision-making can be influenced without their even realising it. For example, *A* pursues a particular course of action because *B* has created an environment such that *A* will follow that course of action without realising that *B* wanted it to occur. In short it is the creation of social conditions that encourage a type of behaviour of which the subjects are not self-consciously aware. Such social control measures are extensive and have been explored by Cohen (1985) in broader society.

To put these in a security context, if a security officer asks someone to leave a shop when s/he does not wish to but they nonetheless do,

THIRD DIMENSION Primary measures	**Creating mentalities to achieve outcomes subconsciously** *Changing malefactors' behaviour* Social measures Deterrence *Refocusing the behaviour of malefactors* Situational measures Design Image and reputation
SECOND DIMENSION Secondary measures	**Presence** Officer presence Product presence
FIRST DIMENSION Tertiary measures	**Effective human element** Verbal questions ◇ Verbal requests (making use of universal and select legal tools) ◇ Verbal threats ◇ Coercion ◇ Call the manager and/or police

Figure 7.1 Lukes's three-dimensional model of power applied to security

that could be considered as an example of the first dimension of power. An example of the second might be where a security officer's mere presence leads someone not to enter the shop although they want to. The third dimension could be illustrated by a particular person not even wanting to enter the shop because subconsciously they have been influenced not to do so; in this case the behaviour of the person has been influenced without there being any observable conflict. How the different dimensions apply to different security measures is further illustrated by Figure 7.1.

The third dimension of power, or 'thought control', according to Lukes could be achieved through control of information, the media and through socialisation. Indeed for Foucault (1977) order was achieved in society through a disciplinary system that encompassed a wide range of mechanisms, other than observable orders, to achieve outcomes. Such subtle controlling mechanisms have been illustrated by Shearing and Stenning (1987) in their study of Disney World and the design, signage and image that are used to secure order. Cohen (1985) has also explored in much greater depth the extensive range of activities that contribute to this 'social control' process at a state level. Many of the measures that are undertaken by the state to engender social control are also pursued by organisations to achieve their own appropriate security outcomes, as well as state social control more broadly impacting upon such locations of governance. Primarily through design, the image cultivated, the rules that are set with sanctions for their breach, and notions of reputation, what Johnston and Shearing (2003) would call a 'mentality' emerges amongst the 'policed', which generally secures the appropriate outcomes. They argue (2003: 29):

> Generally, we act not because we have consciously thought through and adopted a mentality that promotes a particular action, but because we constantly use methods (often embedded in habits) that imply that mentality or practical reasoning. We typically adopt these methods without thinking much about the mentality they imply.

A security system should be predicated on the primary measures of making 'it' – with the 'it' standing for most crimes and other unwanted incidents – never happen. As such the third-dimension power of Lukes and the numerous initiatives that fall within their ambit should provide the primary basis for security systems. While the focus in this chapter is on these third-dimension measures it is nonetheless important to note that the first and second dimensions contribute to the third and that ultimately one has to build a system based upon all three. The design of the security system is of major importance in influencing the risk of targeting (Tilley, 2005a). In setting out some of the many strategies, services and products that can be utilised in security systems in this and the next chapter, we will offer some of the best and most up-to-date examples. It is not possible, neither is it the aim of this book to explore every avenue, but rather it is hoped to illustrate the benefits of a research-led approach to familiar security problems. The burgeoning research and literature under the broad categories of crime prevention, crime reduction, crime science, security and so on gives much food for thought in designing security systems. (Good starting points and triggers for expansion of potential research on different strategies, services and products to enhance security are Gill, 2006 and Tilley, 2005b.)

Changing malefactors' behaviour

The first part of the three-dimensional range of strategies is to seek to change the behaviour of a malefactor (or someone at risk of becoming a malefactor) so that he/she no longer engages in malevolent behaviour. There are two elements to this: the first is what is often termed 'social crime prevention' and the second is deterrence.

Social measures

'Social crime prevention' has been defined as embracing '...almost any program [sic] that can claim to affect the pattern of behaviour, values, and self-discipline of groups seen as having the potential to offend' (Sutton, 1994: 10, cited in Gilling, 1997: 5). Typically social crime prevention measures have focused on countering risk factors that increase the potential to offend, for example, by improving recreational activities/facilities, improving employment opportunities, strengthening families, enhancing the role of schools and tackling drug addiction (Bright, 1997). For most security managers such measures have not been at the forefront of their strategies because they are largely seen as the responsibility of the state. For example, to most supermarket security managers the funding of drug treatment programmes in their locality would seem to be a state responsibility. Further, such programmes are blunt instruments when it comes to protecting a particular node. A drug treatment programme may well stop some of those on the scheme from offending, but it might not affect those who are targeting the supermarket for shoplifting. And such programmes are also more complex than situational measures and take longer to evaluate. A drug treatment scheme can only be properly assessed over several years, and any reduction in offending, and indeed the causes of any change in behaviour might be down to more complex reasons.

Nevertheless despite these concerns social measures do have a part to play in an overall security system. After all it is far better to change a person's behaviour so he/she no longer commits a crime than simply to divert him/her somewhere else in the short term. At the very least organisations could offer financial support to state and charitable organisations providing such services. However, one area of social crime prevention that security managers might look to is developing (more often than not in partnership with statutory bodies) positive recreational activities. In a shopping context, where there might be problems with large numbers of youths hanging around and engaging in anti-social behaviour, such strategies might be more effective than simply banning youths from the area and then having to deploy security officers to enforce the ban (Phillips and Cochrane, 1988).

Riley and Shaw (1985) found that the level of offending is often associated with the amount of time spent with friends away from home. Research such as this, combined with sub-cultural theory has led to the recommendation of numerous strategies that seek to provide constructive activities for (more often than not) young people at risk of engaging in anti-social behaviour (Cohen, 1955). The provision of recreational activities and facilities is one of the more common strategies. The aims of this type of strategy are to provide individuals with something positive to do, rather than just 'hanging around' with the concomitant risks of falling into deviant behaviour. Some of the many strategies that could fall within this category include the provision of youth clubs, sporting activities, building of play areas and so on.

It is frequently youths who are associated with anti-social behaviour in shopping areas. Often it appears that they have nothing better to do than to hang around and there is a risk of them causing a nuisance as well as offending. In Rotterdam a 'drop-in centre' was created to address this problem. Staffed by a youth worker it was a place where young people could congregate to socialise and pursue numerous activities under supervision (Phillips and Cockrane, 1988). This type of approach has been duplicated in a number of areas, notably the Bull Ring shopping centre in Birmingham. Generally such projects are believed to be successful in preventing young people becoming involved in crimes such as shoplifting and causing a nuisance.

Deterrence

We saw in Chapter 3 that many malefactors are rational and weigh up the likelihood of getting caught and the consequences if they do. Deterrence is central to the broader strategies of the criminal justice system in preventing crime. Clearly it does not work all the time, but for some malefactors there is a place for deterrence in the overall security system. This can be established through nodal rules with internal sanctions for their breach, civil law strategies, professional body strategies, criminal prosecutions and finally redress.

In some locations rules can present a considerable challenge, whereas in others they seem natural. A shopping centre needs to maximise footfall in order to ensure turnover is at its highest. Rules, sanctions for breaches and governing structures do not fit well with an enticing environment for shoppers. On the other hand, in a workplace, such as a factory or office complex, such things would be expected. Different approaches can be pursued depending on the location.

The emergence of organisations with their own rules, systems of governance and sanctions for breach has been well documented (Spitzer and Scull, 1977; Shearing and Stenning, 1982; Macauley, 1986). Frequently there are rules on parking, smoking, and the use of bicycles and skateboards enforced by private security at private space locations. Moreover some of these systems actually deal with serious offences, such as theft and fraud, which could also be dealt with by the state criminal justice system (Shearing and Stenning 1982). Sanctions can range from fines or exclusion to termination of employment.

An example of the sophistication of such systems can be found in the research I conducted on Armed Industries (Button, 2007a). There, all employees, contractors and visitors were bound by rules and procedures set out in the employee handbook. The 12-page A5 handbook was issued by the director of human resources and was written in consultation with the trade unions. The rules and procedures in the handbook covered an extensive range of issues from discrimination, expense claims, security issues and safety issues to the use of company telephones. Among the security regulations were a 5 mph speed limit, designated parking zones and a requirement to show company identification (or equivalent). Documentation was given to those driving on site and speed signs were installed that resembled those on public highways. Similarly for parking there were double yellow lines, permits, signs and so on. For identification there were also signs and guidance on the identity cards, and the requirements of this and other issues were also set out in induction training.

Breaches of the rules would be dealt with under the company's 'Code of Industrial Discipline', which was designed to ensure that evidence is gathered, substantiated and properly weighed and that the employee has an opportunity to defend him/herself or be represented by an appropriate person. The document did not set out the role of security in enforcing these rules and it was clear that some would be beyond their remit – such as abuse of telephones. Armed Industries represents one extreme and a shopping complex the other. However, even shopping complexes will still have signs stating that unauthorised parked vehicles will be clamped or towed away and prohibiting skate-boarding or the riding of bicycles.

In some nodes the sanctions go beyond the internal. In Chapter 5 we saw that some organisations use litigation to prevent certain people from coming on to private space by serving them with injunctions. In retailing some companies no longer pursue criminal sanctions and instead undertake civil litigation to recover funds, others pursue both strategies (Bamfield, 1998). In varying degrees, organisations will bring in the police for theft, fraud and related offences, though many are reluctant to do so (Doig, 2006). However, one of the best examples is the NHSCFMS policy of 'parallel sanctions'. This involves a criminal prosecution and disciplinary procedures as well as attempts – where there is a licence to practice involved – to get the malefactor struck off. The NHSCFMS will also seek redress by pursuing civil actions for recompense when the malefaction is fraud or similar crime. For example in 2006 former NHS dentist, Shabir Merchant, was struck off the register of the General Dental Council a few months after having been convicted of defrauding the NHS of almost £200,000, by making claims for work he had not done. He was jailed for 18 months and legal proceedings were pursued to recover the monies lost (BBC News 2006).

These strategies are important in changing the culture of people in a node and encouraging certain types of behaviour. Malefaction might be seen as too great a risk if potential perpetrators think that they might get caught and that the consequences could be loss of job, loss of licence to practice, a criminal record and having to make some payment in redress. Security managers, depending upon the location, can pursue a range of measures to create an environment that makes people less likely to engage in prohibited behaviour. These might include making staff, visitors and the public aware of the rules and sanctions through training, and displaying appropriate publications and signs. People who have breached the rules and been punished can also be used as examples in campaigns to illustrate the consequences of such behaviour.

Refocusing the behaviour of malefactors

The second range of strategies that can be pursued to 'make it never happen' are strategies that aim to refocus the behaviour of malefactors who are unlikely to stop their malevolent behaviour. These strategies are based on the understanding that certain malefactors will not stop offending, so the aim is to reduce their opportunities to offend, move them on to more benign activities or to deflect them elsewhere. Underpinning these types of strategies are situational measures. In Chapter 3 rational choice and routine activity theories were briefly outlined and we considered the importance of opportunity. The implication of these theories is that if the opportunity can be removed, the risk of crime can be reduced. Advocates of this approach, such as Clarke (1980), argue that crime prevention methods should take greater account of particular situations, of the locations in which crimes are committed and of the thinking processes of offenders. A range of security strategies are available to protect property, such as locks, alarms, barriers, CCTV, security guards and so on to make the offender less likely to attack. Thus Clarke and his followers advocate preventing crimes by manipulating the environment in which they are likely to take place – so-called situational crime prevention. There are of course other, broader reasons that help to explain what causes a person to offend – genetics, social problems and the like – which is why some of the social measures were explored earlier in this chapter. These have a role to play, but as Farrell and Pease (2006: 187) argue:

> This does not mean that other social problems should not be tackled – they should – but they are primarily tackled for reasons other than crime prevention. In contrast, influencing the immediate circumstances and context in which crime takes (or does not take) place is the avenue with greatest chance of preventative success ...

A knowledge of situational measures is therefore of central importance to the security manager. Such an understanding can be used to develop an environment that minimises the opportunities for crimes such that malefactors would be deterred from attacking the organisations or would fail if they did. It is worth at this point examining some of the different strategies in a bit more depth.

Situational measures

Situational crime prevention essentially encompasses strategies that seek to reduce the opportunities for crime by increasing the effort of offending, increasing the risks of getting caught, reducing the rewards from committing the crime, reducing provocation and removing excuses (Clarke, 2005). It is often seen as a very useful crime prevention strategy because it is not necessary to understand what motivates the offender only to recognise that some people are criminally motivated. It can also be directed towards preventing specific crimes and protection of specific locations. Clarke (2005) has classified strategies under some 25 techniques under the five broad headings described above. Some of these will be briefly discussed below and the full range of techniques is illustrated in Table 7.1 (those interested in exploring this further should review Clarke, 2005).

Table 7.1 Situational strategies for the security manager

Increasing the effort	Increasing the risks	Reducing the rewards	Reduce provocations	Reduce excuses
1. *Target hardening* Steering locks Bandit screens Tamper-proof seals	6. *Extend guardianship* Leave signs of occupancy Carry cell phone	11. *Conceal targets* Unmarked armoured trucks Off-street parking	16. *Reduce frustrations and stress* Efficient lines Polite service Expanded seating Soothing music/lighting	21. *Set rules* Harassment codes
2. *Access control* Screening staff Entry phones ID badges Card access	7. *Assist natural surveillance* Improved lighting Defensible space design Independent hotlines	12. *Remove targets* Pre-paid cards Remove coin meters	17. *Avoid disputes* Separate seating Reduce crowding Fixed cab fares	22. *Post instructions* 'No parking' 'Private property' 'Extinguish camp fires'
3. *Screen exits* Ticket for exit Electronic merchandise tags	8. *Reduce anonymity* Taxi driver IDs 'How's my driving' deals School uniforms	13. *Identify property* Property marking	18. *Reduce temptation and arousal* Controls on violent pornography Prohibit racial slurs	23. *Alert conscience* Roadside speed display Signatures for declarations 'Shoplifting is stealing'
4. *Deflect offenders* Bus stop placement Street closures	9. *Use place managers* Reward vigilance	14. *Disrupt markets* License street vendors Monitor pawn shops	19. *Neutralise peer pressure* 'Idiots drink and drive' 'Its OK to say no'	24. *Assist compliance* Litter bins Public lavatories

Table 7.1 Situational strategies for the security manager – *continued*

Increasing the effort	Increasing the risks	Reducing the rewards	Reduce provocations	Reduce excuses
5. *Control tools/weapons*	10. *Strengthen formal surveillance*	15. *Deny benefits*	20. *Discourage imitation*	25. *Control drugs and alcohol*
Spray-can sales	CCTV	Ink merchandise tags	Rapid repair vandalism	Breathalysers in bars
Toughened beer glasses	Intruder alarms	Graffiti cleaning	Censor details of modus operandi	Alcohol-free events
	Security guards	Disabling stolen cell phones		

Source: Adapted from Clarke (2005: 46–7).

Strategies that increase the effort of offending essentially make it more difficult for an offender to commit a crime, with the primary aim that of deterring the crime altogether or, failing that, that the offender will instead target elsewhere. The strategies to achieve this include target hardening, such as steering locks, barriers, stronger locks and so on. It also includes measures to establish access control to a specific area, such as identity checks, barriers and the like. It could also involve the vetting and screening of potential staff, contractors and clients. Measures to deflect offenders can also be used, such as closing streets that have become targets of vandalism. Finally the tools used for crimes can be controlled, through what is known as 'controlling facilitators'. Thus in many places the sale of spray paint is restricted to certain groups and in some shopping centres chewing gum is banned from sale, to try and minimise the nuisance of discarded gum.

The next category of strategies, increasing the risk to offenders, covers a wide range of initiatives that make capture more likely. Searches at entrances and exits are a common strategy at airports, night-clubs and libraries, increasing the likelihood that offenders will be caught smuggling weapons, drugs or stealing books respectively. Many shops have EAS where goods are tagged and set off an alarm if they are taken illegally out of the shop. Increasing surveillance through existing staff, security staff, CCTV or simply through natural surveillance is also used to make increased capture a key thought in the minds of potential offenders. For fraud, malpractice and corruption-related issues independent whistle-blowing lines are also an important measure. Staff are more likely to contact an independent line to report suspect behaviour, making deviant behaviour more of a risk. The intelligence thus gained also provides numerous lines of enquiry for security and counter-fraud staff to pursue.

The third category covers strategies that endeavour to make crime less rewarding. One of the most common is to remove the immediate prospect of gain, as has happened with many gas meters, telephone booths and so on, where coins have been replaced by cards. Property can also be marked, which means if it is stolen it is harder to sell and therefore less valuable. Inducements can also be reduced: graffiti artists who take great satisfaction in the public display of their work may be less pleased if it is removed immediately after it has been created. Penalties for breaching rules can also reduce the potential rewards, for example, for illegal parking, as discussed earlier.

The fourth and fifth categories were added by Clarke to the first three, which were set out earlier in a simpler model (Clarke 1992). The fourth covers the reduction of provocations that might trigger deviant behaviour. So factors that might cause stress and thus lead to greater risk of deviant behaviour can be countered – queues can be made to function efficiently, service can be polite and soothing music can be played. Disputes can also be reduced by strategies such as keeping rival groups of fans in separate locations. Certain products are reckoned by some to increase temptation, for example, violent pornography, and controls on this can, it is argued, weaken the potential for arousal and consequent malefaction. Other techniques include neutralising peer pressure through campaigns such as 'idiots drink and drive' and undercutting the tendency to imitation by, for example, censoring information on how crimes could or have been undertaken.

The final category covers techniques that negate excuses. Again, rules can be useful: if it is clear that there is a particular code of conduct in force ignorance of the prohibitions on certain types of behaviour cannot be adduced as an excuse. Clear 'no parking' signs, for instance, are a means to achieving this. Similarly, alerting the conscience of the potential offender is another technique, where for example boards notifying drivers that they are breaking the speed limit may have some impact. Compliance can also be assisted in some cases. So if public urinals are made available, there is no excuse for urinating in the street. Finally drugs and alcohol are often used as excuses, so restricting their use in certain areas can also help.

Thus there is a significant armoury of tools that the security manager can utilise in designing a security system better to protect his/her environment. Many of these also have a part in the physical design of the environment, which will be the subject of the next section, because it deserves consideration in its own right. A visit to any major security exhibition illustrates the wide variety of security products available, ranging in sophistication from metal barriers to very expensive CCTV systems utilising the latest digital technology. Technology is constantly advancing and so are the security products available to the security manager. It is vital that security professionals keep up with such developments through the security press, conferences and exhibitions.

Layers have been central to the enhancement of security from the beginning of time. Schneider (2006) illustrates how layered security was inherent in the building of medieval castles with moats, gates, keeps, multiple walls and so on. Many early communities also established settlements with multiple rings of defence. In Chapter 2 it was demonstrated that in some cases of security failure, one of the contributing factors was the low number of layers. It is common sense that the more layers there are the greater the difficulty in breaching security as the resources necessary to overcome them become greater (Gill, 2005a). Therefore the greater the need to protect an asset the more layers there need to be to achieve that.

Design

The discussion on opportunity also relates to the design of buildings and space. It is, however, worth mentioning some important influences upon design. Jane Jacobs (1961) in her seminal book *The Death and Life of Great American Cities* made some observations on the link between crime in urban areas and public space. She argued that the social controls that existed in small settlements and that controlled people's behaviour were less successful in urban settings where there is greater anonymity. She criticised the planners of her time for failing to recognise that communities are organic and living things and who had as a consequence created unsafe cities that were breeding grounds for crime. She identified various strategies for addressing these problems, such as through the design of buildings so they maximise natural surveillance. This, she argued, would reduce the propensity of some to offend because of the greater likelihood of apprehension, while at the same time enhancing feelings of security.

The work of Jacobs, though little reference was given to her, was taken further forward by Newman's (1972) ideas in 'Defensible Space: People and Design in the American City'. He started with the same critique of modern planning and its impact on the level of social control and natural surveillance. He was particularly critical of high-rise buildings, buildings that look inwards and buildings with many exits. He argued that there were three criteria that increased the rates of crime in residential areas. First, there was anonymity, where neighbours do not know one another. Second, was a lack of natural surveillance, which makes it easier to commit crimes without been seen. Finally, there were numerous exits which make escaping the scene of the crime easy. From this Newman developed his idea of 'defensible space', which is a model for a residential area that creates a physical environment that inhibits crime. There were four elements to defensible space: territoriality, surveillance, image and milieu.

- *Territoriality* The creation of a physical environment so that there are clear demarcations of a zone's ownership.

- *Surveillance* The design of buildings and residential area to maximise surveillance for residents and their agents.
- *Image* The design of an environment so that stigma is minimalised.
- *Milieu* The location of housing projects in safe areas.

Developed alongside Newman's idea of 'defensible space' was the similar crime prevention through environmental design management (CPTEDM) approach. This extends beyond the residential areas explored by Newman to commercial and public organisations/locations. The main elements of the approach are connected to the manipulation of the physical environment so as to reduce the amount of crime and the fear of it, as well as changing the propensity of the physical environment to support criminal behaviour. There are three overlapping strategies that mark the approach. The first is access control, such that access is denied to targets and a perception of risk created for the potential offender. Second, there is natural surveillance where potential targets are designed in order to maximise observation. Finally there is territorial reinforcement, where the importance of defining a territory's ownership is stressed (Oc and Tiesdell, 1997). Boundaries are good examples of relatively simple tools that can influence human behaviour. The use of signage, fences, paving, colour coding and lighting are just some of the strategies that can reinforce ownership (Schneider, 2006).

Designing-out crime has assumed a significant place in the development of new shopping facilities (Oc and Tiesdell, 1997; Crawford, 1998). Indeed Reeve (1998: 74) argues that a mall can be characterised as '...a place of social control in which individuals are ... physically constrained to behave in ways conducive to the ends of production'. Much of this knowledge had been applied at Pleasure Southquay (Button, 2007a), which was designed with clear lines of sight and open spaces to maximise natural surveillance. The site was very well lit, particularly the car park. The mixed retail and leisure usage meant that people would be attracted to the centre at night as well, minimising the chance of an 'empty' centre and the feelings of insecurity that might create (Poyser, 2003).

Most security managers inherit an environment that they have had little involvement in designing. However, the principles of CPTEDM can often be used to enhance existing environments. As an example, if a university campus is experiencing a spate of men exposing themselves to female students in grounds where there is extensive shrubbery, possible initiatives might include removing or pruning the shrubbery to increase natural surveillance, better lighting for the hours of darkness, introduction of CCTV or even, if possible, the closure of the path or redirection. As another example, if an office complex is troubled by trespassers regularly roaming through the grounds measures could be introduced to demarcate ownership, such as signs, barriers, fences, changes in colour or even a line.[1] In addition to CPTEDM there have also been numerous initiatives around 'secured by design' (SBD) where buildings and some products are assessed for their criminogenic influences. For example research has shown that housing designed with SBD experiences 30 per cent less crime than buildings which do not incorporate it (Farrell and Pease, 2006). Such initiatives have also been very useful in addressing crime in car parks (Park, 2005).

Drawing upon the broader literature on designing-out crime Ekblom (2005) identifies four principles:

- *Designing products to be inherently secure.* For example designing café furniture so that there is a slot underneath the table to put a bag, thus making less likely the placement of bags on a table or on the floor or hanging off the chair – where the risk of theft is higher.
- *Adding on security products.* For example adding an alarm to a car.
- *Restricting the resources of the offender.* For example limiting knowledge of potential vulnerabilities in a product.
- *Securing the situation.* Using broader situational measures in which the potential target is located.

Image and reputation

Eco (1976) has demonstrated that there are certain 'signals' that have a disproportionate effect on a person's perception of risk and behaviour. The ideas from this research have been developed and applied to crime by Innes and Fielding (2002). They argue that 'signal crimes' – incidents that because of the way they are interpreted by people act as warnings about the risks in a specific area – have become recognised as important in the development of reassurance policing (which can be similarly applied to the principle of enhancing security). It is argued that if certain signals have a disproportionate impact on the behaviour of individuals, tackling these could have a significant impact on feelings of security. At its most extreme this can be illustrated by the abduction of a young child, which, although extremely rare, invariably leads to parents becoming increasingly concerned at the risk of the abduction of their children and modifying their behaviour as a consequence. More typically, however, more commonplace crimes and examples of anti-social behaviour can have an impact that goes beyond the actual victim into the wider community. As Innes and Fielding (2002: 5) argue:

> ... the commission of a crime can lead law abiding community members to perceive they are at greater risk of victimization. As a result of which they withdraw from using public space, thus decreasing levels of informal social control in an area. Concomitantly, the commission of the crime and the lack of an effective response to it, can act as a signal to potential motivated delinquents and criminals that this is an area where there is decreased chance of being apprehended or challenged if you engage in delinquent or criminal behaviour.

The principles of the above are more commonly articulated through Wilson and Kelling's (1982) widely-accepted argument that broken windows lead to more broken windows. More empirically-based evidence, which has supported this, has been provided by Skogan (1990), who established the idea of 'decay spirals'. He argued that failure to deal with trivial acts of anti-social behaviour could lead to an escalation in their number and to more serious acts of crime. Thus it is very important that an organisation avoids presenting an image characterised to any degree by signs of decay and anti-social behaviour such as vandalism, graffiti, litter and so on.

As signal events can have a disproportionate impact on an individual's sense of security and behaviour – which affects both the public and malefactors – so tackling them and sending out the signals (and therefore the image) that they will be dealt with can have a disproportionate impact in a positive way. Thus an organisation's image and reputation in dealing with crime and anti-social behaviour is a very important part of the security system. At Pleasure Southquay cleanliness was considered as of central importance, with a team of cleaners of equal size to the security force. Problems with cleanliness were immediately dealt with by the cleaning team. Consequently anyone, or any activity, that compromised the cleanliness of the centre would look out of place (Button, 2007a). As Shaftoe and Read (2005: 255) argue:

> A high quality, cared for environment will encourage respect for that environment and its users ... Conversely harsh, fortified and neglected environments may reinforce fear and actual risk. There

is evidence to suggest brutal surroundings may provoke brutal behaviour...

The importance of image, however, goes beyond ensuring that there are no signs of decay. There are a large number of nodes in which certain behaviours and types of clientele are wanted. Top-class hotels often create an environment that obliges one to behave in a certain way and that may also discourage many from even considering entering. Similar strategies are used as appropriate for other locations, such as shopping and leisure complexes. Therefore image and the use of marketing strategies to achieve the required image have become increasingly important tools in the security armoury (Harvey, 1990; Reeve, 1998). Raco (2003: 1870) has argued that, '...creating safe, aesthetically pleasing public spaces requires the removal of "social pollutants" – those individuals and groups whose (co)presence may threaten the perceived and aesthetic quality of an urban space'. Research has also noted how higher expectations of behaviour can be created through what has been called 'domestication by Cappuccino' or the creation of a more bourgeois environment where different norms of behaviour are expected and tolerated (Zukin, 1995; Atkinson, 2003; Massey, 2005). Indeed Lees (1997: 339) found such an effect in the opening of Vancouver's new library,

> ... homeless or street people and other so-called social deviants are not specifically excluded by library policy, the fortress style architecture, security consciousness, and middle class ambience (bourgeois playground) of the library, in general, does nothing to attract them or make them feel at home.

Again we can illustrate this with the example of Pleasure Southquay (Button, 2007a). There, image and marketing were incredibly important and were central to the overall security system. The promotional literature for Pleasure Southquay sought to create an image as an exclusive and unmissable shopping location. Some of the prominent headlines on the literature at the time included, 'Destination Unmissable', 'The Ultimate Destination for a Unique Christmas Experience', 'Ultimate Shopping Destination', 'Ultimate Lifestyle Destination' and 'Ultimate Leisure Waterfront Destination'. This literature contained pictures focusing upon yachts and sailing – thus referencing a very exclusive and expensive pastime. Exclusive 'designer' outlets such as 'Gap', 'Ralph Lauren', 'Tommy Hilfiger' and 'Paul Smith' were promoted. Literature also focused upon 'restaurants' and 'cafés' and dining 'al fresco' and on entertainment based on 'contemporary artists' and comedy; a style of entertainment distinctly different from the traditional working-class pubs across the road, which were frequented by some of those who were not targeted by Pleasure Southquay.

By focusing on exclusivity and the 'designer' end of the high street, and on distinctly 'upmarket' leisure activities the promotional material created an image that appealed to consumers with – or aspiring to – lifestyles based on the display of consumer wealth. Many of the local youths, unemployed and low paid, would not consider such a place as appropriate for their own shopping or pursuit of leisure because of the image and the expense. Thus the promulgation of such an image would deter some simply because of cost, and others would be likely to feel uncomfortable in such exclusive surroundings.

The image and reputation of the security staff is also an important part of this strategy. A reputation for high quality security can do much to send out signals that will deter attacks on a particular place or organisation. As the case study on El Al below reveals, its reputation for top quality security acts as a security layer in its own right, deterring terrorist organisations from attempting attacks. In contrast, if security has a reputation for slackness, sleeping on the job, incompetence and so on, that image may well send out signals encouraging attack.

It is also important to note that the appropriate image and reputation of a node can be achieved by more than just the environment and what people perceive when they visit or discuss it with those who have. Bowers and Johnson (2005) have identified a wide range of publicity tools that can be used to prevent crimes, from offender-targeted strategies, such as personal communications with known offenders and strategies to encourage public action, such as crime prevention advice, to crime prevention intervention publicity, such as the release of information to the media about successes or an impending crackdown. Indeed Laycock (1991) found evidence of the way in which intense local and national press coverage led to a reduction in the rate of burglary.

Case study: El Al

Aviation has proved to be one of the most attractive targets for Islamic terrorists and the country of Israel is probably the most hated by those groups. This might lead to the assumption that El Al, Israel's national airline would be a regular target of terrorists. However, since 1968 there has been no successful attack on El Al and the few attempts that have been made have all been thwarted.[2] El Al has one of the best security systems in the world and in the view of Wilkinson (2006) it is the only airline up to the challenges posed by modern terrorism.

In 1970 when hijackers attempted to storm the cockpit of an El Al flight the pilot sent the plane into a dive which knocked the terrorists off their feet, enabling the air marshals to overpower one and kill the other (Shuman, 2001). The next major attempt on El Al came in 1979 when an Islamic terrorist befriended a German recently released from jail and asked if he would traffic drugs to Israel. He agreed and the pre-flight screening showed nothing unusual, but in the profiling interview he was unable to explain how he purchased his ticket in Switzerland but was flying from Germany. A further search revealed that he was carrying explosives rather than drugs (Business Week Online, 2003). In 1986 another attempt was foiled when a pregnant Irish woman attempted to board an El Al flight at London Heathrow carrying explosives, which were discovered. She had been duped by her boyfriend (and the father of the child) into carrying the bomb, of which she was unaware. In 2002 at the passenger counter of El Al at Los Angeles airport a gunman opened fire randomly at a queue of passengers, killing two. The gunman was shot dead by an El Al security guard (BBC News, 2002c).

Not surprisingly El Al does not give a great deal of information out about its security. From various sources, however, it is possible to piece together the strategies they use to address the risk of terrorist attacks against them. First of all their security begins long before the actual flight. The El Al flight schedule is regularly changed to make it difficult for terrorists to plan ahead. All passengers are vetted against El Al databases to assess any potential risks as soon as names are booked on flights, as are the purchasers of tickets. Any potential doubts about a passenger are logged for special attention in the screening procedures. On the approach to airports in Israel there are check-points that passengers must pass (Shuman, 2001; Walt, 2001; Business Week Online, 2003).

Once passengers reach the airport and check in they undergo further tests. Passengers are divided between Jews, non-Jews and Arabs. The latter two groups are often interviewed by up to three separate profilers who ask extensive questions about the purpose of the flight, the luggage and so on. As illustrated by the case of the German drug smuggler above, such techniques can expose flaws in plans which enable a potential attack to be averted. All luggage is screened and subjected to a pressure test (the pressure will detonate any pressure-triggered bombs) as well as a hand search. All bags must be reconciled with a passenger on board before the flight is allowed to leave.

Once on board there are sky-marshals sitting anonymously amongst the passengers who are highly trained ex-military personnel. The pilots are also usually ex-air force pilots trained in various techniques to use should the plane be hijacked or targeted with missiles. Indeed all security staff involved with El Al are highly trained and motivated, with many from ex-military backgrounds. The pilots are also protected by double doors. All El Al staff members are subjected to regular tests and face disciplinary action if they fail (Shuman, 2001; Walt, 2001; Business Week Online, 2003).

If the layers of security applied by El Al are compared with those faced by the hijackers on 11 September the strengths of the former are further demonstrated. The 9/11 hijackers were unlikely even to have considered attacking El Al, because the strength of their security regime is too good and would pose too many risks, and there were far easier targets to choose (see Table 7.2).

El Al's security is an example of a security system designed to make it never happen. Its reputation is such that terrorist groups are unlikely to consider targeting El Al because any attempt would be doomed to failure. El Al's security measures, however, are expensive and time consuming (passengers must check in three hours before departure) and it is debatable whether the strategy used for an organisation that has 40 flights a day – in the USA alone there are 35,000 flights daily – could be widely applied. Nevertheless there are elements of the model

Table 7.2 Comparing the layers of security between El Al and 9/11 American levels of security

El Al	American aviation security on 9/11
Excellent reputation	Poor reputation
Disruption of schedules	Pre-screening system (focused upon bomb)
Investigation of future passengers	Screening of person, luggage and hand luggage
Screening of person, hold luggage and hand baggage	
Interview with up to three security profilers	
Air marshals	
Double doors	
Security trained pilots	

that could be utilised with little inconvenience to improve security in a variety of areas.

- *Targeted security*: focus greater attention and layers on the greater risks.
- *Multiple layers*: the system has multiple layers to minimise the chance of success should some parts of the system fail.
- *Highly motivated and trained security personnel*: well trained and motivated staff.
- *Regular tests*: tests to ensure security is effective with consequences for failure.

Conclusion

This chapter began with an explanation of how Lukes's three-dimensional conception of power can be applied to security. It then went on to examine strategies that fall under the third dimension and, to some degree, the second, that can make it never happen or at least minimise the chances. These strategies were divided into two parts: those that seek to change malefactors' behaviour and those that aim to refocus the behaviour of malefactors. Under each a very wide range of ideas were examined, many of which have been shown to be effective. As such this chapter has showcased some of the many different 'tools' from a range of disciplines that can be used in enhancing the effectiveness of security (that is, in reducing the risks of crime and other deviant acts). The case study of El Al served to demonstrate how an effective security system can indeed make it never happen. In the next chapter the main elements in the first and second-dimensions of a security system will be explored: the human element. The focus will be on security officers, but there are other personnel who undertake similar duties and technology also has a role to play.

Notes

1 The next time you are lying on a sandy beach draw a line in the sand around yourself and watch how many people who see the line deviate around it.
2 Although in 1996 a Hezbollah agent did manage to smuggle a kilogram of explosive through Zurich and Ben Gurion airports and only failed to achieve his aim by blowing himself up in his hotel (Wilkinson, 2006: 125).

8
Making the Last Resort Count

Introduction

In the last chapter we outlined the three dimensions to an effective security system and looked in depth at the third-dimension strategies. In this chapter the main component of the first and second-dimension strategies will be explored: the human element. (Although it must be remembered that the human element also contributes to the third-dimension strategies, by reputation for example.) The importance of the human element has often been underestimated in the design of security systems (Lane, 2001). Drawing upon appropriate research this chapter will show how the presence (second dimension) and interventions (first dimension) of the human element can be enhanced. It must also be remembered that security officers are not always the human element and last resort, there are other personnel that also undertake this role, such as police officers, community support officers, wardens, military personnel and in some instances other staff members of an organisation. Technology may also be the last resort. Nevertheless security officers will be the focus of this chapter, though many of the comments could also relate to some of the other 'security' personnel.

The chapter will begin by exploring strategies that can enhance performance before the security officer even puts on a uniform. These include the recruitment and selection of staff, their working hours and their training (although this issue, because of its link to regulation, will be explored in the next chapter in greater depth). The chapter then looks at the toolbox of a security officer and assesses how its use can be maximised, before exploring systems that can ensure maximisation of quality. A case study of Intelligarde Security draws the chapter to a close by offering an exemplar of a high quality last resort.

Recruitment and selection

It is not the aim of this section to become too embroiled in the debates over the most effective recruitment and selection techniques. Rather it will focus upon some key themes and issues which should be borne in mind when recruiting and selecting security officers. The first strategy that should be pursued is to set pay and conditions that attract the appropriate quality personnel for the job. The old adages that 'you get what you pay for', and 'if you pay peanuts you get monkeys' could not be more appropriate for security officers. The fact that many organisations are willing to trust the protection of often extremely valuable assets to personnel on the minimum wage beggars belief! Uncompetitive pay and conditions mean security staff are recruited from the bottom end of the market, making poor quality recruits more likely, leading to potentially less competent security staff more open to deviant influences, and increasing the probability of higher labour turnover and, as a consequence, the size of the pool of people having 'inside' knowledge of the place being protected. Therefore the first strategy in recruitment and selection should be to set pay and conditions at rates in the local market that are not at the bottom of the market for the skill of the job. This should attract better quality applicants and minimise labour turnover.

Once the job has been advertised the next stage is the recruitment and selection. There are a wide variety of different methods that can enhance the traditional interview, for example, role plays and psychometric tests. The practice of Intelligarde – the case study at the end of this chapter – where potential recruits 'ride along' with existing staff, to assess how they react to different situations also seems admirable (Rigakos, 2002). However, probably the most important part of the selection process is the vetting or screening of the potential recruits. As chapters 2 and 4 showed there are many examples of security officers engaging in corrupt behaviour.

As we saw in Chapter 5 most countries have a licensing system for security officers which involves a character check and where a licence is only issued if an applicant has no criminal convictions that bar entrance to the occupation. In some systems there are provisions which enable the rehabilitation of offenders by granting licences if the offences are not serious or if a specified period of time has passed since the conviction for a particular type of offence. For example in the UK, the SIA mandates a period of two years for minor offences and five years for more serious offences where there have been no further convictions, cautions or warnings. There is a very complex matrix of offences which, depending on the length of time since conviction and the seriousness of the crime, inform the decision of the SIA (SIA, n.d.a). The SIA website criminal record indicator can be used to assess whether particular offences would make a person ineligible for a licence.[1] So for example it would be possible for someone with a conviction for arson under s 1(3) of the Criminal Damage Act 1971 who was sentenced to prison but who has been free of warnings, cautions and further convictions for over five years since the end of the sentence to be eligible for a licence. For many clients, entrusting their premises to a security guard at night with such a conviction might seem too much of a risk.

It is also important to remember that many offences committed by employees do not result in criminal convictions, just termination of employment (Hollinger and Davis, 2006). But would a warehouse storing electrical equipment want a security guard who was sacked (but not prosecuted) from their last job for pilfering stock? The SIA does not investigate the family or associations of an applicant, although it is quite possible to imagine a scenario in which a potential candidate with no convictions might be married to a woman who is the sister of a convicted armed robber. Such a scenario might pose too much of a risk to a cash-in-transit company. Therefore the SIA only provides a very basic level of vetting – as is the case with most other licensing systems in other countries. For some security companies this is the only vetting undertaken. Indeed in November 2007 a scandal broke concerning the vetting of security staff by the SIA when it was discovered that 'thousands' of illegal immigrants could have been given licences (BBC News, 2007c). However, the SIA's duty is to check identity, criminal records and competency; the duty to check the immigration status is a statutory responsibility for the employer. Therefore this scandal really exposed the poor additional vetting by security companies. Many better quality companies engage in more detailed vetting to at least BS 7858 (this includes proof of identity, record of education and employment, character references, criminal convictions, any bankruptcy and court judgements and statement of authorisation and acknowledgement: misrepresentation or failure to disclose are grounds for dismissal) (British Standards Institution, 2006).

Vetting of staff is seen as a major strategy in the prevention of inappropriate behaviour by prospective security staff and covers a wide range of different strategies. The foundation of most vetting strategies, as set out in BS 7858, is an application form (or CV) and the verification of the information supplied. Some studies have shown that significant numbers of people lie and embellish CVs. One study found that one in three people questioned made false claims about their qualifications, experience, interests and past jobs. It also found that 20 per cent of 1000 workers surveyed had exaggerated on their CV (BBC News, 2001b). Therefore checking the qualifications claimed by applicants, writing to previous employers to confirm previous jobs and periods of employment and checking references forms the foundation of most vetting strategies. However, it is interesting to note that the same study that found that applicants made false claims also found that a third of 350 managers surveyed did not check applicants' CVs because it was too time consuming. The growth of identity theft has also been increasingly noted, with estimated costs to the British economy in the region of £1.3 billion to £1.7 billion (Cabinet Office, 2002; Duffin et al., 2006). Therefore the need to check documents to confirm a person is who they say they are is also important. These types of checks confirm evidence of the honesty of an applicant.

There are some sectors where further specialist checks might take place. For example a pharmaceutical company that conducts experiments on animals might ask a specialist screening company to investigate whether an applicant has any links to animal rights groups. In the USA it is also common for applicants to take drug tests, integrity tests and even lie detection tests (Fischer and Green, 1998). For more senior positions, more extensive vetting often takes place, usually through specialist screening or private investigation companies. Indeed there is a strong market in companies offering the full range of vetting services. Typical of the services

Table 8.1   Carratu's employee vetting services

Level 1	Level 2	Level 3
Confirmation of address and other contact details	Level 1 + Property ownership search	Usually for the very highest appointments and includes
Credit referencing and history including bankruptcy search	Search of judicial and legal data	Level 1 + 2 + Additional relevant research to confirm the probity of the applicant
Confirmation of employment history and references	Company directorship search	
Validation of relevant educational qualifications and membership of professional bodies	Full media/database search	

Source: Carratu (n.d.).

offered by many firms engaged in this market are the three levels of vetting offered by Carratu International (Table 8.1). Many of these are also applied by organisations to contractors and vendors.

Appropriate working hours

Chapter 4 discussed the often long and irregular working hours of security officers. Here, the consequences of such practices are considered with suggestions from research of what can be done to counter their effects. There has been a great deal of research on issues relating to working hours including night work, shift-work, long hours, sleep deprivation and so on. The hundreds of studies that have been conducted often provide contradictory results. There is also debate on the definitions of concepts such as fatigue and sleepiness and on how to measure performance (Monk, 1991). This is not the place to engage in those debates, although there are some common themes that begin to emerge from the literature on this subject and that can provide clues to enhancing the effectiveness of security officers.

Before the extensive research on the impact of different working hours is considered it is worth looking at the most obvious consequence of the long working hours demanded of many security officers: that they are more likely to sleep on duty. This happens in either an organised way, where officers deliberately seek out a place where they can go to sleep, or more randomly, when they are simply so tired that they fall asleep wherever they are. There is a huge amount of anecdotal evidence for this. Type 'sleeping security guard' into the search engines of Google images or You Tube and dozens of pictures and pieces of footage are found. Perhaps a good illustration of this comes from Ross McLeod's highly entertaining book documenting his creation of Intelligarde in Canada (McLeod, 2002). In the book he describes his early recruitment to a security firm in Toronto:

> I said to them, 'can you find me a site where I can sleep through the night?' they said, 'yup,' and hired me on the spot. They sent me to Mark's Work Warehouse to get the obligatory blue shirt and dark pants and gave me one or two badges. And that is how, in 1982, I started in the security industry at $3.75 per hour.
>
> (McLeod, 2002: 20)

Studies specifically focusing on security officers and the impact of their working hours have been very rare and I am only aware of one, by

Alfredsson et al. (1991), on Swedish security guards. The research compared 197 security officers with a representative sample of 1769 males. This found that security guards had 2–3 times higher occurrence of sleep disorders. They suffered a higher frequency of fatigue, insomnia, nervous problems and depression.

Police officers, who often conduct similar functions to some security officers operating within comparable conditions (though often with better working hours) have been the subject of more research, particularly in the USA. These have shown the devastating impact of fatigue on the health and performance of police officers. Fatigue is largely caused in police officers by long working hours arising from double shifts (to cover sickness and for overtime), shift-working more generally and from moonlighting (Vila et al., 2000). These lead to loss of sleep and disruption of the circadian rhythms which culminate in fatigue. The consequences of fatigue were impaired performance and worsened mood.

Levels of fatigue amongst police officers were found to be a major barrier to the introduction of more effective policing. Fatigue impairs the formation of sound judgements, restricts potential choices of decision and finally induces poor responses through irritability, ultimately lowering the quality of decision-making (Vila et al., 2000). Vila et al. (2000) found their sample of officers routinely worked more hours over one, two and seven day periods than would be legal for lorry drivers or those working in a nuclear facility. Using the FIT pupilometry test they found 6.1 per cent of officers were highly impaired, with impairment equivalent to a 0.10 alcohol concentration and 19 per cent of officers were impaired (in the UK the driving limit for alcohol is 0.08). Additionally 41 per cent of officers reported having clinical sleep pathologies. Below we will focus on some specific issues related to the impact of different types of problematical working hours on performance and health.

Irregular shifts

It is a common finding in the literature on working hours that irregular shift patterns, particularly night-shifts, lead to fatigue and sleepiness, which impairs performance (Åkerstedt, 1991). There are a number of studies that illustrate this. Bjerner et al. (1955) found that mistakes in the readings from a gas meter were at a peak during the night-shift with a secondary peak in the afternoon over a 20-year period. Browne (1949) demonstrated that telephone operators took longer to connect calls at night and Hildebrandt et al. (1974) found train drivers failing to operate their safety devices more often at night. Simulations of pilots flying at night have shown a deterioration in performance equivalent to moderate alcohol consumption (Klein et al., 1970). Consider also some major disasters, such as Chernobyl, Three Mile Island and the Challenger Space Shuttle disaster, all of which happened or were caused in the early hours of the morning (during the night-shift) or emanated from decisions made at that time (Åkerstedt, 1991).

Working over 50 hours per week and/or long shifts

There is also much research to illustrate that working long hours, 50 hours per week or more, impacts on performance. In one factory in the UK working hours were reduced from 53 to 48. Production levels stayed the same and the rise in wages per hour was offset by reduced fuel costs (Spurgeon et al., 1997). Studies on munitions workers during the war found that a reduction in the hours of work by between 7 and 20 hours a week resulted in no reduction in output and in some cases seemed to promote an increase. Output increased further with the introduction of scheduled rest breaks (Spurgeon et al., 1997). As well as effects on performance, longer hours increased the likelihood of mistakes and, consequently, of accidents. There is also evidence that longer shifts, particularly where work is physically and mentally demanding, lead to a decline in performance, particularly towards the end of a shift and even more so if it is a night-shift (Waterhouse et al., 1992).

Impact upon mental and physical health

There is also a great deal of evidence of the impact of irregular and long hours on physical and mental health. Irregular and long working hours disrupt the circadian rhythms causing a wide range of health concerns.

An overview of research on the impact of shift-work found that prolonged working in such patterns increased the mortality rates (Waterhouse et al., 1992). The same overview concluded that there was evidence of a link between long-term shift-working and cardiovascular diseases, digestive disorders and ulcers. There is evidence that longer working hours lead to increased stress which leads to gastrointestinal disorders, musculoskeletal disorders, weakening of the immune system, and psychosomatic complaints that culminate in more sick leave and further reduced performance (Spurgeon et al., 1997). In one study of patients who had experienced coronary failure under the age of 40, 71 per cent had, over a prolonged period, worked days and evenings or more than 60 hours per week (Russek and Zohman, 1958). Research from California found that amongst men under the age of 44 who worked in non-physical jobs for over 48 hours per week there was a greater risk of coronary heart disease (Spurgeon et al., 1997). For women working over 46 hours per week links have been made to higher rates of abortion (McDonald et al., 1988).

Waterhouse et al. (1992) in a review of the literature on long-term shift-work found some studies suggesting increased nervousness, anxiety and depression and overall a tendency towards general malaise compared to day-workers. Other studies have found more specific examples of problems. Spurgeon et al. (1997), utilising a range of examples, and a standard measure of psychiatric status, found a study of urban bus drivers demonstrating higher ratings of mental illness than the normal population. In Japan, where working hours are generally much longer, a study of factory workers found that alongside other factors such as poorer nutrition and lack of physical exercise those working more than 9 hours per day were in greater psychological distress. These stresses often lead to 'maladaptive behaviours' where excessive smoking, drinking and drug-taking are used as a coping mechanism. A study in Australia of coach drivers found that long driving hours were the best predictor of stimulant abuse. Given that these types of working hours apply to security guards it is likely that many of these attributes also apply. Indeed, one small-scale study of security guards, entitled 'The Psychotic Patient as Security Guard', concluded, 'The frequent admission of security guards to an inpatient psychiatric service raises serious social policy concerns' (Silva et al., 1993: 1439).

None of this research on working long hours and/or shift-work includes the inevitable impact on family, social life and so on. The impact on the quality of life is a factor that has not really been developed in the research, but which surely also ultimately has an impact on performance and health.

Lessons from the research

Addressing the effects of working hours upon security officers requires a difficult balancing act of squaring performance, health and social issues. For what maximises performance might not be the most social hours and what is best for the well-being of an officer might not maximise performance. Additionally some working patterns may make the pursuit of secondary employment (moonlighting) more feasible, undermining any positives. Bearing these in mind the literature would suggest some of the following strategies as likely to enhance performance.

The evidence suggests that working for more than 50 hours a week is detrimental to health, well-being and performance and that fewer working hours than this (including any additional employment) would seem most appropriate in all cases (Spurgeon et al., 1997). If duties are physically and/or mentally demanding shorter shifts are likely to prove more effective than longer shifts (8 hours rather than 12) (Waterhouse et al., 1992). For those on shift-work including night-shifts there are a number of strategies that can be pursued to enhance effectiveness. Harma et al. (1986) have shown that increasing physical fitness can greatly reduce fatigue and increase alertness on the night-shift. Åkerstedt (1991) has advocated that sleepiness on night-shifts could be minimised through short night-shifts, rotating the pattern clockwise (mornings, afternoons, nights; rather than nights, afternoons, mornings), by having the night-shift at the end of a pattern, and through slow rotation.

The environment in which the security officer works can also be enhanced to minimise sleepiness. Åkerstedt (1991) advocates avoiding sedentary tasks and striving for roles that are not boring. He goes on to argue for an environment that should also be cool, with good lighting and high levels of irregular noise. On a personal level Åkerstedt also argues for strategic napping before a shift, particularly a night-shift.

The security officer's toolbox

After the recruitment and selection and the setting of appropriate working hours the natural next strategy would be appropriate training. However, Chapter 9 will examine the principles of effective minimum standards of training and before we can advocate what security officers need to be trained in, it is first important to assess the 'tools' in their box. Therefore this section will outline the toolbox of the security officer and training will be picked up in more depth in Chapter 9.

The toolbox of the security officer provides for a potentially potent last resort. Box 8.1 outlines the main tools which will be considered in detail below. Symbols of authority such as uniforms and badges can enhance compliance and reduce the incidence of verbal abuse, threats of violence and actual assault. There are also extensive legal tools available to a security officer, although not all personnel are fully aware of these. There are also important linguistic tools of persuasion open to security officers that can be used to maximise compliance. Other tools of knowledge relate to issues such as understanding non-verbal behaviour, which could prove vital to a security officer. There are also specialised tools that could be used to enhance the effectiveness of

Box 8.1   The security officer's tool box

Symbols of authority (uniform, badges etc.)
Legal tools and knowledge
Linguistic tools
Specialist knowledge
Special tools (weapons, handcuffs, night sights etc.)
The body

---

security officers, such as lethal weapons, and more defensive weapons and equipment such as handcuffs and body armour or tools such as night-sights. Finally there is the body itself and in particular knowledge of control and restraint, martial arts and so on.

Symbols of authority

A significant strategy for enhancing the effectiveness of security personnel in terms of the second-dimension (visibility of staff leading to deflections of offenders) and first-dimension (compliance with staff requests) strategies are based upon the principles of people's general obedience to authority. The most notable research on this were the experiments of Stanley Milgram. In these experiments members of the public were invited to the university laboratory to participate in an experiment, for which they were paid a fee. Their task was to ask questions of another participant (in reality an actor) and every time the other participant got a question wrong they were to administer an electric shock. In fact there was no electric shock and the actor only pretended to experience pain by screaming. Close to the participant was an authority figure dressed in a white coat and overseeing the experiment. When the participant questioned the experiment the authority figure in the white coat would tell them to carry on. There were numerous results from these experiments depending upon some of the different permutations found. One of the most striking findings was that nearly two-thirds of participants in some scenarios were prepared to administer lethal electric shocks. However perhaps most useful was how the experiments showed that the intervention of the authority figure in various scenarios generally induced compliance. Milgram showed that people are generally obedient to authority (see Milgram, 1975).

This provides very useful knowledge for the security manager in creating symbols of authority in order to secure compliance. Indeed it has been argued by Loader (1997) that the police derive significant authority from symbolic images based on uniforms, equipment, rituals and so on. Similarly private security personnel also derive varying degrees of authority from the images they present (Stenning and Shearing, 1979; Cooke, 2005). Further to these findings and ideas there has been research illustrating the benefits of uniforms in terms of gaining compliance. Most private security personnel wearing police/military style uniforms drive in officially marked vehicles, carry 'official' identity badges, and in some countries carry weapons (De Waard, 1993, 1999). Nearly all security personnel represent in some way a corporate body or organisation that is seen by some to have power and authority which adheres to a degree to private security staff as their agents. These all combine to give private security staff a degree of authority. As Davis (1957: 22, cited in Becker, 1974: 439) states,

> The mere presence of a uniformed individual contributes a psychological condition of great significance to the average mind. Over the period of many years, the wearer of the uniform has represented a leader, designated and recognised by government bodies. This association has been attached to the form of distinctive wearing apparel.

There is research specifically on uniforms that illustrates why these are so important to security staff. Bickman (1974) undertaking experiments using a civilian, a milkman and a guard was able to show that the latter – through the uniform worn – secured much greater compliance than the others vis-à-vis the public. Bickman used three scenarios in which an actor dressed as each character in turn asked a stranger to pick up a paper bag, asked a stranger to lend a dime to another actor who needed it for a parking meter and finally, at a bus stop, strangers were asked to stand on the other side of the sign. Table 8.2 shows the different levels of compliance in the three different scenarios.

Table 8.2 Percentage of subjects complying in three different scenarios in Bickman's experiment

	Paper bag	Dime	Bus stop
Civilian	36	33	20
Milkman	64	57	21
Guard	82	89	56

Source: Bickman (1974: 51).

Bickman's research demonstrated that the 'guard' secured significantly more compliance in all three scenarios than the milkman or civilian. Further research on this subject by Bushman (1984) also found that compliance increased with the perceived authority, where a person dressed as a fireman received more positive responses to requests made to the public than a man dressed in a business suit or as a 'bum'. Bushman also found that compliance was greater amongst older subjects.

The apparent benefits of a uniform inevitably lead to the debate that frequently occurs in locations such as shopping malls, leisure facilities and the like of whether to have guards in more traditional security uniforms or dressed more casually in slacks and blazers. There have been numerous studies around this theme and some of the evidence is conflicting. However, in a study by Harris (1974) it was found that individuals dressed in low status clothing received significantly more verbal, non-verbal and overall aggression than those dressed in higher status clothing. In another study by Mauro (1984) examining data from a police department that had changed from traditional paramilitary dress to slacks and then back, no significant change in the number of assaults on police officers was found. However in experiments utilising members of the public and their assessment of uniforms he found that:

> ... wearing the traditional uniform, officers were rated as fairer, more honest, more 'good', more active, more helpful, faster, and (from the pretest) more physically fit, more intelligent, and better able to defend themselves, compared to when wearing a blazer. Clearly, the blazer is not demonstrably superior to traditional uniforms; indeed, it appears to be significantly inferior in a number of respects.
>
> (Mauro, 1984: 54)

The positive benefits of compliance, however, need to be balanced against the potential risks of the impact on an individual of wearing a uniform. The wearing of a uniform has been shown to induce changes in behaviour. The (in)famous prison experiments of Haney et al. (1973) showed that 'normal' individuals in uniform could perpetrate sadistic acts, and Zimbardo (1969) has also argued that uniforms may deindividuate the wearer, reducing their moral responsibility for actions and increasing the potential for aggressive behaviour.

All of this provides numerous (sometimes conflicting and caution-ary) lessons for the security manager seeking to enhance compliance amongst those they are responsible for policing. First is the importance of having a more paramilitary style authoritative uniform rather than a blazer. Second is the importance of the display of symbols of authority such as identity badges, flags and so on. Linked to this is the impor-tance of weapons, which can be used by security staff in some coun-tries. The famous quote that 'power flows out of the barrel of a gun' provides further illustration of the symbols of authority. If guns, trun-cheons and other non-lethal weapons can be used their display can also send important messages of authority. Clearly, however, staff need to be appropriately trained in their use, and this is explored below.

Legal tools and knowledge

Security officers have a wide range of legal tools available to them in most jurisdictions. These are based on citizens' rights to arrest, use force and so on; from operating on private property; and from securing areas where those visiting are under a contract, particularly a contract of employment (see Stenning and Shearing, 1979; Sarre and Prenzler, 2005; Button, 2007a). There are also some security officers who have been given special legal powers, such as those at airports or in courts (see Jason-Lloyd, 2003). Based on where they operate I have developed three models of security officer power (Button, 2007a). The 'basic secur-ity officer' draws only upon citizens' rights, such as to make an arrest and to use reasonable force. The 'semi-empowered security officer', who is based on private property can, additionally, draw on the rights of private property owners to exclude and remove individuals from private property, search those entering and leaving (with consent) and enforce other conditions such as parking regulations. Finally there is the most powerful officer who is the 'complete empowered security officer' who operates in a space where those entering are in a contrac-tual relationship, most usually an employment contract. The powers that can be derived from this include search and enforcing conditions, as well as the powers of the 'semi-empowered'. Table 8.3 sets out the different models with the powers attached.

Despite the extensive legal tools that are available to security officers, the limited research that has been conducted on them has suggested poor levels of confidence in their use and in some cases a poor under-standing of the tools that are available. The first study of this kind was the 1971 Rand Study on American security officers (Kakalik and Wildhorn, 1971d). Among some of the more disturbing findings of this

Table 8.3 Models of security officer power

Model	Powers available
Basic security officer	Universal legal tools • To ask • To arrest • To use reasonable force to prevent a crime
Semi-empowered security officer	Select legal tools: property based (reasonable) • To exclude entrance to private property • To remove from private property • To enforce conditions on private property • To search person on condition of entrance
Complete empowered security officer	Select legal tools: property based (arbitrary) and employment and/or contractually based • To enforce conditions on private property or other area • To search person on condition of entrance • To search person on exit from private property

study were the facts that 18 per cent of respondents said they did not know what their legal powers were and a further 23 per cent were unsure of them. Less than 50 per cent of respondents knew that their arrest powers were the same as an ordinary citizen and 6 per cent thought they held the same powers as a police officer. Only 22 per cent knew under what conditions an arrest for a felony was legal. Thirty-one per cent of respondents thought it was a crime if someone called them a 'pig'! Six per cent of respondents would have been prepared to use deadly force to protect private property and 19 per cent thought that as long as an arrest was made in good faith the security officer could not be subject to civil action.

In a study of security officers in South Korea 46 per cent did not know their powers of arrest and 34 per cent were somewhat unsure of them. Seventy-two per cent were either somewhat unsure of or did not know their powers of search and 71 per cent were somewhat unsure of or did not know their powers to use force (Button and Park, forthcoming). In research on a small group of English security officers the results were better (Button, 2007a). One-fifth were 'somewhat unsure' on their powers to use force, over a quarter didn't know what the legal grounds of arrest were, 10 per cent thought they had the same powers as police officers and some officers made worrying mistakes in case studies of scenarios where they might have to use their legal tools.

The overall analysis of the security officers' responses to 13 questions testing their knowledge (Button 2007a) revealed that, on average, they would make one major error, one minor error and answer 11 satisfactorily (for median as well, but for mode 0, 0 and 11 respectively). More of a concern was that further analysis revealed that a very small group of officers accounted for a disproportionate number of errors. Indeed overall, three of the 49 officers accounted for nearly a quarter (13) of the 55 major errors and six of the 49 officers accounted for a third (18) of the 54 minor errors. Thus, a small minority had a very poor knowledge of their legal tools. The standards and training at the two case study sites were typical of (if not better than) the average in the British security industry, which suggests that the broader picture would be much the same. Indeed one covert ethnographic study of a major UK security firm (which at the time complied with all relevant British Standards on security) found the induction training officer spoon-feeding the answers to candidates and virtually guaranteeing that no one would fail (Adu-Boakye, 2002). These findings illustrate the need for better training in the use of such tools as well as the management of the process.

Linguistic tools

There is much research illustrating how police officers are able to avoid using their legal powers by exploiting various linguistic tools. Police officers might ask 'What have you got in your bag?' rather than 'Open your bag so I can search it', asking a question rather than forcing an individual to do something (Dixon, 1997). The threat of the use of legal powers, arrest, search warrants and so on is another strategy used by police officers. Police officers might also secure consent by exploiting a suspect's ignorance. Perhaps one of the most significant factors in enabling police officers to secure consent is the implication that refusal to comply with police requests is tantamount to an admission of guilt. Thus, if suspects are asked to empty their pockets, if their house can be searched or if they will come to the police station to 'help with enquiries', in many cases they will comply for fear of appearing guilty if they don't.

These same strategies can be and are used by security officers. At Pleasure Southquay security guards often used questions rather than making demands. One officer said that he would ask a suspected shoplifter, 'Would you mind coming with me?' Another officer told me, 'You stop him and explain who you are. At no time do you say "you are under arrest" as you get into the "every thing you say will be..." Nine times out of ten once they're stopped they generally come back anyway' (Button, 2007a: 89).

The other strategy used by security officers both at Pleasure South-quay and at Armed Industries was threats to secure compliance. One security officer at Pleasure Southquay said that he would threaten to call the police. Others evoked a lesser sanction of threatening to call or report the person to management and raising the prospect that they could be sacked if they refused to cooperate. There was little evidence of bamboozlement (confusing people), but this could also be a tactic for security officers.

If security officers are properly trained, so that they know their legal tools, and are well-grounded in the tactical skills of asking questions, utilizing threats and, potentially, bamboozling suspects, they have the potential to be very effective. Such tactics, however, are rarely covered or rehearsed in depth in security officer training.

Other knowledge

There is a vast amount of other knowledge that can be of use to a security guard. In the UK, conflict resolution has been mandated within the 22.5 hours training requirement to satisfy part of the competency standards for a security guard. This subject and the broader area of non-verbal behaviour are of considerable use to security guards who have to deal with the public. For example, such knowledge can offer insights on whether a person is telling the truth, is nervous or is becoming aggressive. Many security guards have responsibilities relating to fire prevention and emergency procedures and health and safety. Therefore knowledge of hazards, hazardous substances, different causes of fire and best practice responses could all prove useful. A knowledge of first aid would also be valuable. These are just a few of the more prominent examples.

Special physical tools

Depending upon the jurisdiction there are a wide range of special tools available to security officers to enhance their performance. Table 8.4 lists the most commonly available. At the top of the spectrum are lethal weapons. These weapons are designed to kill and they include rifles, handguns and shotguns. In some countries such weapons can be legally carried by security officers, either through special legislation for the security industry or through citizens' rights to carry arms. There are armed security guards in the USA, Australia, Canada and South Africa and, in Europe, in Austria, Belgium, France, Germany, Spain and

Table 8.4   Specialist tools available to security officers

Lethal weapons	Non-lethal weapons	Defensive tools	Utility tools
Rifles	Truncheons	Handcuffs	Radios
Handguns	Batons	Protective vests	Torches
Shotguns	Tazer guns	Helmets	CCTV
	Pepper spray		Camcorders
	Dogs		Night-sights

Switzerland (De Waard, 1993; O'Conner et al., 2008). The importance of appropriate recruitment, selection and training of security officers who carry lethal weapons was graphically illustrated by the American Rand report which cited numerous instances of the misuse of weapons, including recklessly discharging firearms and shooting bystanders. However, probably the most graphic was the case of a boy swimming in Lake Barcroft, North Virginia, who ignored a security guard's instruction to stop and was shot dead (Kakalik and Wildhorn, 1971d: 140). The report is dated but there are still periodic exposés of security guards misusing weapons throughout the world. For example in Bratislava prior to an England v. Slovakia football match in 2002, in a bar where the staff were having difficulty removing around 60 England fans the manager called in a security firm to help. In the ensuing melee (with unarmed England fans) 16 bullets were discharged and two fans were hit (BBC News, 2002d). In most jurisdictions the discharge of weapons in the pursuit of public order would be regarded as inappropriate and such cases illustrate the dangers of arming security officers. Clearly there are dangerous assignments throughout the world in which such weapons are required, but they should only be provided after an appropriate risk assessment and the recruitment, selection and training of staff capable of using them responsibly.

Non-lethal weapons are the next level of specialist tools, although they can often prove lethal in practice. Among these are truncheons, batons, tazer guns, pepper sprays and guard dogs. In most countries where lethal weapons are allowed so are most non-lethal weapons. Elsewhere the situation is mixed. For example in the UK, security guards are not allowed non-lethal weapons – although guard dogs are allowed. Even guards protecting cash and bullion in armoured vehicles do not carry non-lethal weapons. In South Korea, general security officers can carry non-lethal weapons such as batons, and firearms can be carried by another category of officer, special security officers (Button et al., 2006).

The next level of specialist tools are defensive tools such as hand-cuffs, protective vests and helmets. These are common to almost all jurisdictions, although handcuffs are not that common in the UK and their use is debated (Ralph, 2004). Finally there are tools that provide the foundations for undertaking security work. These include radios, torches and specialist tools such as night-sights. In some situations – such as protests – where allegations of excessive force are often made, security officers also carry cameras to video their colleagues' and pro-testers' actions (Button and John, 2002). Clearly it is essential that officers are trained in the use of all these types of equipment.

The body

In certain aspects of security work the body is of vital importance. For example, for door supervisors in bars or clubs it is essential to have an appropriate physical presence (Hobbs et al., 2003). Equally – given the level of physical confrontation faced by door supervisors – such staff should be trained in control and restraint and martial arts. Most security officers will experience less physical conflict than door super-visors, although there are some areas of security work where they do occur. Security officers working in retail environments where arrests are common should clearly be trained in control and restraint. Others working in environments where there is much alcohol should also be trained in such skills and there is also a case for some defensive martial arts skills.

Creating an environment to maximise effectiveness

We have looked at the physical tools available to security officers to help them undertake their roles more competently. In this section we examine other strategies – simulations, tests and the more con-troversial policy of going undercover – that can be used to create an environment which will maximise the effectiveness of security officers.

Simulations

Simulations and games of varying levels of sophistication are frequently used in training for the military, public sector workers and by some corporations. In the security context, many managers undergo such scenarios for preparations relating to crises and disasters (Borodzicz, 2005). For security officers the simulations and gaming that they are most likely to undertake will be related to the evacuation of buildings.

Simulations and games, however, have considerable potential in the training of security officers for different situations. Confrontations with members of the public, arrests, searches and patrols all provide the potential for difficult situations that could be better prepared for through simulations. This is a vital tool for the enhancing of security which is substantially underutilised.

Tests

Traditionally the central mechanism for ensuring effective practice for security officers has centred around guard control systems. These are usually physical devices in which the guard places a key at strategic points or scans a magnetic strip or barcode during a patrol. These produce data which enable management to assess the timings and route of patrols. Such devices can be sabotaged by guards who can move the keys and strips. There is also the burdensome task of analysing the data. Such systems also provide no qualitative data. Though their use should not be discounted, there are more effective means to test the quality of security.

Chapter 2 discussed the extensive use of tests in aviation security and the generally poor performance revealed thereby. Exposés by journalists of poor security also act as tests (again some of those were explored in Chapter 2). In computer security, vulnerability testing is well established as a means of testing security systems in which experts – often known as ethical hackers – test systems and produce reports that highlight weaknesses and areas for improvement (Andress, 2002). The benefits of using former criminals to test security, was considered in Chapter 3 vis-à-vis 'The Heist'. In security environments where it is vital that security staff stay alert, tests are an essential part of strategies to ensure that effectiveness is maintained. In any organisation where large numbers of searches or checks on identity to secure access are required, tests should be of central importance to maintaining quality. In Chapter 4 one of the consequences of the poor culture of security officers was demonstrated to be a very low level of commitment to checking passes or conducting searches. If the company had used regular tests to identify such bad practice and had invested in retraining (and, where appropriate, had disciplined malpractice) such negative behaviour would have been rarer. At Pleasure Southquay, tests were conducted using 'mystery shoppers'. These are members of the public recruited to sample a wide variety of services and write a report on their experiences. At Pleasure Southquay they also reported on the security staff, although the main focus was on enhancing the officers' customer service skills. So, for instance, an officer might be asked where the nearest toilets were. Fear that a member of the public might be a mystery shopper undoubtedly kept the officers on their toes! There are some companies in the UK that offer testing services, such as the Perpetuity Group and the Security Watchdog. As part of its services the latter organisation will seek to test security by breaching the access controls of a client. This they argue will,

> ...test the effectiveness of the security operation of the end-user's premises. The Watchdog will try to gain entry into the premises, which are usually secured by a manned guarding contractor. These Audits are without notice and have a positive long-term consequence in that the security operation is kept alert at all times. Access Audits can occur at any time in the day or night. The end-user receives a confidential Report the following day.
>
> (The Security Watchdog, 2007)

If the security of an organisation is to be maximised, regular testing by experienced practitioners keeps security alert and contributes to maximising its effectiveness.[2]

Going undercover

The most extreme measure to ensure effectiveness is to go undercover – or what could be described as covert ethnography. This involves sending a person to join a security team whose real purpose is to report on what they are actually doing. Such practices are common in research, journalism and in the police. Ordinary research-oriented ethnography raises many ethical challenges relating, among other things, to the safety of the researcher, the risk of becoming involved in criminal activities and the potential harm to the 'researched' (Murphy and Dingwall, 2007). Going undercover raises the same and wider concerns. If discovered, such action might provoke angry reactions from employees, compromising industrial relations, and in a unionised environment it might lead to industrial action. Nevertheless if there are major concerns over the activities of security staff it can secure valuable data. For example, in one covert ethnographic study the researcher uncovered a number of dubious practices: he was asked to lie by management and state that an arrested shoplifter had also been racist to ensure a harsher sentence, there was a complete lack of supervision of his activities and staff who attended court were not paid (Adu-Boakye, 2002). In an episode of 'Tonight with Trevor McDonald' a journalist went undercover with the security staff at Birmingham airport and exposed staff sleeping on duty, inadequate searches and worst of all an officer reading the paper while supposed to be screening luggage for explosives (BBC News, 2007a). Appropriately skilled undercover personnel are usually only found through private investigators, which means that it is not a cheap option (researchers would be bound by issues of confidentiality, anonymity and the potential harm to subjects involved, making their use unethical). Ultimately this is an extreme option only to be pursued if all other measures have failed and there are deep concerns over aspects of security.

Effective relationships

It is also important to note that in many cases security offices are able to secure the help of the police or some other statutory body as a last resort. Indeed there are many 'watchman' type security officers who would call the police in preference to intervening in many situations. The relationships between different groups of security officers and between security officers and the police, community support officers, wardens and other relevant bodies are therefore very important. Stenning (1989) has demonstrated how the relationship between the police and private security can go through six stages.

1. *Denial.* Police officers refuse to acknowledge that private security officers are a legitimate topic of discussion.
2. *Grudging recognition.* The relationship moves on to grudging recognition accompanied by denigration, where the increased role of private security can no longer be ignored, but recognition of its existence is combined with a belief that security officers are undertaking lesser activities, such as property protection.
3. *Competition and open hostility.* Under the third stage the relationship moves to competition and open hostility where the growth of private security poses a threat to the police's claimed monopoly over policing, but at the same time is recognised as a source of employment in retirement and a means of providing policing services under stretched budgets.
4. *Calls for greater controls.* The fourth stage of the relationship moves to calls for greater controls on the industry which has now become a necessary evil or legitimate component of policing.
5. *Active partnership.* The fifth stage is active partnership where it is realised that private security has a significant role in policing.

6. *Equal partnership.* Finally Stenning identified a stage that had not yet been reached in Canada – where the research was conducted – at the time of writing, that of equal partnership between police and private security.

The status of these relationships may vary between local, regional and national nodal levels. Clearly a positive and at least active or equal partnership is essential at all levels. At a nodal level, in most circumstances, the aim of the relationship should be – at the very least – to enable mutual help and assistance whenever required. At Pleasure Southquay I found a very strong relationship verging on equal partnership between the police and private security, supporting one another in difficult public order situations, as well as sharing intelligence and even in some cases powers (Button, 2007a). At a higher level Project Griffin offers another example of a strong relationship between private security and the police. Under this scheme the police provide training to security personnel on various counter-terrorism issues, have a communications network to alert the security sector to the latest intelligence and, should a major incident occur, have a system in place to utilise the security sector in specific areas (Metropolitan Police, 2007).

It is, then, possible for the public and private sector to work together positively. From the perspective of the security manager it is vital to build such relationships so that in extreme circumstances the support of other agencies can be relied upon. And from a police perspective a strong relationship with private security provides a significant resource for them to draw upon when they are overstretched.

An effective last resort: the parapolice – Intelligarde Security Canada

As an example of best practice for security officers I am going to use a somewhat controversial case. Intelligarde is openly marketed as 'private law enforcement' or as an alternative to the police, something that many operating in both the security industry and the police would not want. However, there are a number of reasons why Intelligarde offers a model for effective security officers.

Intelligarde was founded in 1982 in Toronto, Canada, by a former sociology professor, Ross McLeod. It has established itself as an innovative company, with a law enforcement orientation and a high standard parapolice culture. Originally operating in some very lawless public housing estates in Toronto, it now has a wider range of clients.

Central to the philosophy of the company is 'pushing the envelope' in the use of legal tools, with security staff actively using legal powers. The company has been the subject of extensive research by Rigakos (2002) and the owner has also written a highly readable history and philosophy of the company (McLeod, 2002).

Starting with recruitment Intelligarde pursues innovative and model policies. First, to attract better applicants Intelligarde paid above the market rates (McLeod, 2002). The company also actively marketed itself to college students wishing to pursue careers in the police and comparable occupations. Indeed a strong aspect of the company culture is that it is a transitory employer for those seeking such work, although it is more than happy also to offer careers. Those who get past the initial assessment of the application form must go for a 'ride along', where they are attached to actual serving officers, which enables them to decide if the job is for them and allows the company to assess the candidate in the actual job setting (Rigakos, 2002). During the 'ride along' applicants are also tested on their ability to write a report of what they have experienced, which is an important part of the job. Successful candidates undergo training that covers report writing, communication, the law and self-defence. The basic training consists of three days in the classroom and two days field training. There are, however, extensive further training courses that staff can pursue depending upon their role. Specialist courses are available to staff covering a range of subjects from trespass law, interviewing and the use of the tactical baton to cultural issues (Intelligarde, 2007). The culture of preparation for a career in the police means that many recruits are keen to do as much training and to get as much experience as possible.

The toolbox of Intelligarde officers is capacious. As a very deliberate strategy the officers were given high authority uniforms in the formal colours of black and white (McLeod, 2002). Matching livery is used for their vehicles and other equipment. They are also properly trained in their legal tools and actively encouraged to use them. As Rigakos (2002: 77) observed of those attending training on legal powers: '...talk of private powers captivates the candidates. Many of them can hardly believe how much authority they will be able to exert on private property. They are told, "As long as you can justify it, make the arrest."'

Indeed McLeod (2002: 67) boasts of 30,000 arrests by his company in 20 years with no deaths or long-term disability of detainees in custody and no officers convicted of malpractice: what he rightly describes as, 'A world class record for any police service, public or private.' The ongoing training of officers covers other useful knowledge, such as crisis intervention, self-defence and martial arts. Intelligarde officers also have extensive specialist tools available to help maximise their effectiveness, including handguns, batons, guard dogs, horses, handcuffs, body armour, camcorders and radios.

There is also an extensive system of control to ensure that officers are doing what they are supposed to and to provide data on what they are doing. They carry a small Deister Gun that reads barcodes. On patrols, strategic points have bar codes that officers scan to record their presence. Officers also carry barcodes for arrests and calls for support, which also provides useful information to management. Surveillance of staff is made through the radio system and a requirement to report all incidents centrally. Close scrutiny is made of the officers' equipment, uniform and presentation and any deviation is punished. The company has built a strong hierarchical structure based on rank and prestigious roles, such as the Problem Resolution Officer (PRO), considered one of the elite within Intelligarde (Rigakos, 2002). It is also worth mentioning that in undertaking its functions in different areas the company generally tries to pursue a three-dimensional approach. Surveillance, tackling minor infringements, image, intelligence analysis and design of security are among the tactics built into the overall strategy to maximise their impact.

The law enforcement orientation of the company is not suitable for all security contexts. Nevertheless there are many principles embedded in Intelligarde that mirror the strategies that have been highlighted throughout this chapter. In that context it offers a beacon of best practice in maximising the last resort in the security system, the human element.

Conclusion

This chapter has explored the most important part of the security system in the second and third dimensions to achieving effective security: the human element. It began by exploring some of the strategies that can be used to enhance security before a security officer even puts on a uniform. That is, recruitment and selection strategies and utilising effective working hours. The chapter then went on to assess the wide range of tools available to the security officer and how these can be utilised to maximise effectiveness. Among the tools that can be used to ensure that the quality of security can be maintained are simulations, tests and, most controversially, going undercover.

Finally, Intelligarde was introduced as one of the best examples of an effective last resort. Underpinning a lot of the issues discussed in this chapter is the need for appropriate training. This will be explored in the next chapter together with the connected and wider issue of regulation, the need to create a more effective professional infrastructure and the subject of tackling security inequity.

Notes

1 See http://www.the-sia.org.uk/home/licensing/door_supervision/wizard/cri_ intro.htm.
2 When the author was working as a security guard for the West Midlands Police there was a manager who would regularly try and breach security, sometimes in the middle of the night. The security guards were extremely alert to this potential risk.

9
Rebuilding the Foundations of Security

Introduction

This chapter will set out an agenda of reforms that can rebuild the foundations of security. As with the rest of the book this discussion will centre around the UK, although much also applies to other countries. The chapter will begin by considering ways in which regulation can be strengthened to improve the quality of the private security industry. It will then look at the creation of an appropriate professional infrastructure that could produce a more professional means of doing security. In particular, we argue here for a new model of security management – 'Security Risk Management'. Finally the chapter explores the more contentious issue of security inequity and sets out a framework utilising 'security unions' to tackle this.

Introducing effective regulation

It is important to recall that security services and equipment are often a grudge cost and that purchasers will often take the cheapest option (George and Button, 2000). In unregulated markets this provides an incentive to lower costs and standards of security in order to win business. If one supplier is charging X , the supplier who can offer X minus Y will in many cases win the contract, even if the reduction has been achieved by lowering the number of days training, the quality of supervision, wages and so on. This creates a 'Dutch Auction' where the next supplier then wins the contract by offering X minus Y minus Z and lowering standards further. It is for this and other reasons (see Button and George, 2006) that many countries have introduced statutory regulation to set minimum standards below which the market will not fall. The effectiveness of these different regulatory systems varies considerably. George and Button (1997) and Button and George (2006) identify the key characteristics and most effective models of the regulation of private security. I do not intend to go over these models in depth again here, but the broad principles of them are briefly set out below.

The first key principle should be that regulation applies to the wider private security industry. There are many sectors in the private security industry, covering, among others, occupations such as security officers, security managers, door supervisors, close protection officers, security managers, private investigators, security consultants, installers of security equipment and locksmiths, employed either contractually or in-house. For the benefits of regulation to be fully felt it should apply to all of these. Covering the whole of the industry also reduces the risks of leaving loopholes which the unscrupulous could exploit. As Chapter 5 revealed, in the UK the failure to implement regulation of 'security consultants' has already led to some criminals exploiting this weakness to establish a role in the private security industry.

The second major principle of regulation is that it should be deep. Regulatory systems vary a great deal in terms both of their depth of regulation and the most effective mandatory standards for firms and employees. For employees, the best regimes mandate standards beyond character. Standards should also cover minimum levels of competence and staff training. They should also apply to the firm as well as the employee. Thus the licensing system should be for employees *and* firms. Only with licensing of both can regulation make the private security industry truly accountable. Putting an employee's or a firm's licence at stake – with some form of sanction for non-compliance – provides a much greater incentive to comply. These two criteria underpin the 'comprehensive wide' model of regulation in which regulations apply to the wider private security industry, to both firms and employees, and where the standards set cover quality-enhancing criteria.

Another important element is the degree of compliance towards the regulation. There are significant differences in the culture of compliance and degree of enforcement in different jurisdictions and amongst different occupations (Hyde, 2003; Lister et al., 2001). Therefore, what on paper might seem like a 'comprehensive wide' system might turn out to be much less effective when analysed in terms of the degree of compliance. Unfortunately there has not been a great deal of research on compliance with regulation. In the absence of such research all that can be done is to indicate whether compliance is 'high' or 'low', where in the latter there is significant evidence of non-compliance with the legislation. Clearly the aspiration should be a system that maximises compliance.

In terms of responsibility there are three categories. First of all there are *monopoly* systems, where all parts of the private security industry are regulated by one body. Second there are *divided* systems which can be further split between *functional* and *territorial*. As the name suggests in a *divided* system there is more than one organisation responsible for regulation. In the *functional* version, different organisations take on responsibility for regulating different parts of the private security industry and in the *territorial* model responsibility is shared for the same sectors but certain responsibilities are devolved to another body in a defined geographical location. The most efficient regulatory model is *monopoly*, as this minimises organisational disputes over responsibilities and different interpretations over rules and processes. Linked to this is the degree of independence of the regulator. Pure self-regulation, where the industry is given the power to regulate itself, is counter-productive, as the regulation can be captured and serve the interests of the regulated rather than the broader public interest. Similarly giving the responsibility to the police, who have a competitive interest in the private security industry, is also counter-productive, as they may begin to regulate the industry with a view to inhibiting its ability to perform certain police functions. It is also important to note that regulating industries is not the core business of the police.

This then leads naturally to independent regulation, where the regulatory body is neutral towards the regulated industry and where the broader public interest is at the forefront of practice. The problematic element here is the possibility that regulations might emerge which are impractical, frustrate the industry and are over bureaucratic. In Chapter 5 the issue of the mentality of regulation was also raised and it was suggested that Ayres and Braithwaite's (1992) model of responsive regulation offered a way forward. In this model the presence of interest groups is enshrined within the regulatory process, but they do not have ultimate control. The appropriate interest groups are built into the regulatory process, which means that they are able to help shape the regulations, but not to determine them. The benefits of this model are not only that it is likely to result in the creation of more practical regulations, but that such regulations are more likely to be complied with. The regulatory body should therefore formally integrate the key groups representing the regulated industry, the purchasers and the public into the regulatory formulation process for consultation. The British regulator, the SIA, has a wide-ranging stakeholder engagement strategy that encompasses a broad range of different strategies, including the use of the media, the internet, research, surveys and periodic meetings (SIA, 2007), but there is no forum that brings all these groups together and enables two-way debate between the regulator and the stakeholders. A consultative council with the key stakeholders represented could do this. In Finland, for example, the regulator is a government department in the Ministry of the Interior, but a key part of the regulatory structure is an advisory board composed of representatives of the security industry (employers and employees), the business world and consumers of security (Section 51-2 of the Private Security Services Act 282/2002). Its responsibilities include:

- To further cooperation between the authorities and the security service sector;
- To define the general guidelines for the security sector;
- To monitor and promote international cooperation in the security sector;
- To monitor developments in the security sector, security training and research, and guidelines and information concerning the sector;
- To devise initiatives on the security sector;
- To issue statements concerning the security sector when so requested by the Ministry of Interior.

> (Section 52 of the Private Security Services Act 282/2002
> – Unofficial translation)

This advisory board can bring a positive influence to bear on the regulator by ensuring that restrictive regulations are avoided, without actually dominating it. From the regulator's point of view it also provides invaluable advice. This provides an example of very good regulatory practice.

A model regulatory system

A model regulatory system should, therefore, encompass the following principles. Regulation should be applied to the wider private security industry, should be comprehensive and seek to enhance the quality of the industry. The regulatory system should be a monopoly regulator, either a government department or a quasi-autonomous body at arm's length from government. The regulator should be given appropriate resources to undertake effective enforcement. It should also integrate stakeholders in the regulatory process through various consultative measures centred upon a consultative council that brings the regulator and stakeholders together to discuss regulatory issues.

Perhaps the best example of a system that fits these principles is the regulatory system created under the Private Security Authority (PSA) of the Republic of Ireland. The responsibility for regulation lies with the PSA which is an arm's length body appointed by the government. Unlike the British SIA, however, the board of the PSA incorporates considerable industry expertise. Of the nine board members, five come directly from the private security industry (two employers, two employees and one member from the Security Institute of Ireland), with the rest either from the police, the government or independent representatives. In my view the industry should not have a majority on the board, but it does mean that there is expertise at the 'top table'. The PSA also licenses employees and firms and it extends its influence over the wider private security industry. Sectors licensed include:

- door supervisors
- suppliers or installers of security equipment
- private investigators
- security consultants
- security guards
- providers of protected forms of transport
- locksmiths
- suppliers or installers of safes

The legislation also extends to in-house security guards and door supervisors. The licensing system covers a range of requirements, but the most important are the quality enhancing standards. Contractors are required to meet the appropriate industry standards: I.S.999:2004 for Door Supervisor and Security Guarding Contractors, EN50131and SR 40 for Contractors installing Intruder Alarms and I.S.228 and SR 41 for Alarm Receiving Centres. For individual applicants for licences evidence of having passed the appropriate training course is required. In the case of security guards this is either: Basic Guarding Skills FETAC Level 4 Minor Award (L12407 or C10266) or Security Industry Awareness FETAC Level 4 Minor award (C10164) or equivalent (Private Security Authority, n.d.).

So in the UK system the following reforms would need to be introduced. First, regulation should be extended to the wider private security industry so that the portfolio of regulated occupations would be security officers (static, CCTV and CVIT), security managers, door supervisors, close protection officers, private investigators, security consultants, installers of security equipment and locksmiths (contract and in-house). Second, the Approved Contractors Scheme should be made compulsory for all firms. Third, a Private Security Consultative Council should be formally established, representing the key interests in the private security industry and other stakeholders and given a formal role in consultation on new proposals and existing operations. Fourth, minimum standards of training for security managers should be established and a higher threshold created for employees. The latter two are very important and these will be considered below. For employees the discussion will focus on security officers, because this is the most relevant occupation in this book, but the same principles of enhancing training need to be considered for other employees as well (for door supervisors, security consultants, private investigators and so on).

The training of security managers

As we saw in Chapter 4 the training and education of security managers and their commitment to the job is poor in the UK. The inadequacy of the regulatory system is confirmed by the absence of any competency standards or special training standards for security managers. Security managers are leaders and are key agents in promoting cultural change within the sector. They should provide the model for inspiring junior security officers to aspire to managerial positions. Instead, many security managers in the UK have done no special security management training and many have come from outside the security industry, particularly from the police and the military. This sets completely the wrong example to the employees lower down the ladder. Chapter 5 discussed the example of the NHS where regulation has begun with minimum standards of training for security managers. Under this scheme, managers must do a training course that lasts about 5 weeks and equates to Level 3 on NQF and 40 Level C credits towards the first year of an undergraduate degree. This level of training should be mandatory (either as special training or equivalent through accredited prior or experiential learning) for a security manager and should be the first rung on a ladder of higher levels of voluntary training. These should be linked to a professional infrastructure and grades of membership which will be explored later in the chapter. The syllabus of the proposed base-level award is set out in Box 9.1 and has been influenced by the old IISec (now SyI) basic training courses, the reformed version delivered by Perpetuity Training and the Accredited Security Management Specialist course as well as the ideas generated in this book.

Box 9.1 Model syllabus for entry level security management training

General management
Health, safety and fire
The management of risk
Principles of security management
Research and evidence-based practice
Law and regulation
Specialist option (retail, logistics, information, hospital, emergency and disaster management, computer etc.)

The syllabus would provide for a basic introduction to the principles of management, such as planning, performance, motivation, supervision and so on, as well as introducing some of the latest managerial thinking. It would also cover general issues relating to fire and accident prevention and some of the legislation governing these issues. The principles of risk management, from conducting a risk assessment through evaluating risks and identifying strategies to dealing with incidents, would also need to be included, as explored in Chapter 6. The principles of security management, including the most appropriate security solutions for different scenarios would also be covered, as considered in Chapters 7 and 8. It is also vitally important for security managers to understand the importance of evidence-based practice, that is, the selection of proven strategies and continuous monitoring of their effectiveness. Law and regulation are increasingly important items in the in-tray of a security manager and knowledge of the key parts of regulatory law, criminal law and civil law relevant to security are also essential. Finally the specific context within which particular security managers function also needs to be taken into account. Box 9.1 outlines some of the areas in which specialist training might be appropriate. This is just a brief flavour of what should be in a model syllabus and is not intended as a definitive list. It does, however, provide a skeleton on which the full body of a course could be built.

However, this should be the bare minimum and all security managers should be encouraged and incentivised to secure higher level awards. A true profession is dominated by graduates and the degrees and masters or equivalent awards available should become the norm. There is also a case for greater fusion of the latest management thinking with the most up-to-date security wisdom. Gill et al. (2007) have argued for more MBA courses to include security modules – something that is very rare. Indeed most security management academic courses in the UK are rooted in either the criminology or management disciplines with only partial recognition of the interconnectedness of the disciplines.[1]

There is a good case to be made for new awards that more closely integrate the latest management thinking taught in MBAs to the latest security thinking often found in criminology and crime science courses. Indeed at my own university (Portsmouth) at the time of writing (May 2008) we have been considering such a course under the banner of a taught MPhil or completely new MSA – Masters in Security Administration. This is an issue that will be developed further later in this chapter.

The training of a security officer

It is common sense that the more training a person undergoes the better they are likely to be at a task. Chapter 4 highlighted the limited standards in most countries for the training of security officers. This section will set out the core areas of training that a security officer should undertake in order to practice. It is influenced by the toolbox outlined earlier as well as the current UK industry standard and the European Vocational Training Manual for Basic Guarding (SIA, n.d.b; and Spaninks et al., 1999). In South Korea legislation has created two types of security officer: the 'special security officer' and 'general security officer', with different levels of training and rights (Button et al., 2006). This provides a model to build upon, for the different working context of security officers requires two broad streams of training under the separate headings of General Security Officer (GSO) and Public Security Officer (PSO). These will now be outlined (see also Table 9.1).

Most security officers working in the private security industry do not regularly confront physical danger or have to use force in arresting a suspect. The vast majority of security officers therefore, should undergo a 'General Security Officer' standard of training. This should include the role of the security officer and understanding the principles of security, and should cover subjects such as access control, patrolling and basic principles of situational crime prevention. There is also a need to understand the basic principles of law and the legal tools available. This should include powers of arrest, powers to use force, search, criminal law and procedures and litigation. Communication skills are essential and the most important of these are writing reports, customer skills, persuasion skills and radio procedures. Many security guards, no matter where they work, are likely, at some point, to find themselves in situations of conflict, and knowledge on recognising non-verbal

Table 9.1   A model training syllabus and structure for security guards

Security officer type	Syllabus
General Security Officer (GSO)	The Role of the Security Guard and Principles of Security Legal Tools and Law Communication Skills Conflict Resolution Fire Prevention Health and Safety Diversity and Human Rights Well-Being Level 2
Public Security Officer (PSO)	GSO + Advanced Legal Tools Control and Restraint Court Skills First Aid Level 2
General/Public Security Officer Supervisor (G/PSOSv)	GSO or PSO + Supervisory Skills Level 3

behaviour and resolving conflict situations is essential. Most security guards are required to undertake basic fire prevention and health and safety tasks and being able to recognise hazards and knowing what to do with certain types of fire is also essential. The problems of racism in the police have been well documented (Macpherson, 1999; Reiner, 2000) and while there is much less evidence for racist conduct among security officers, it is essential that any potential for discrimination is reduced and that officers are fully aware of the importance of respecting human rights. This should be mandatory. Finally, given the hazards of the job, not just in relation to potentially violent offenders, but also the dangers to health and well-being as a result of shift-work and irregular hours, security officers should also be trained in strategies to enhance their well-being, such as when to sleep, when and what to eat, how to recognise the signs of health problems and so on.

In the UK, many of these subjects are already included in the knowledge-based training to secure competency. However the current length of training is only 22½ hours, or three days, which does not seem adequate to deal with the wide range of essential issues. Just to get an understanding of the law and legal tools open to an officer would probably take the average person at least a day. However, mandating hundreds of hours training as other countries have sought to do is also not the most efficient way forward, for it does not recognise the very different abilities of security officers. Driving licences in the UK are issued not on the basis of a period of practice, but on a standard of competency that might take some candidates a few dozen hours to achieve and others hundreds. There are also increasingly good mechanisms of delivery through distance and E-learning which make contact hours meaningless. Therefore a better solution would be to mandate a standard that an officer must achieve within a period of time and that would be tested through a formal exam, written assessment and demonstration of skills on the job. This should be achieved with a week of foundation training in the classroom before they can begin working, combined with follow-up training and support which could be delivered in the classroom, on the job, by distance or via E-learning or through a combination of these methods. This should culminate in an assessment leading to a Level 2 award on the NQF. Security officers who have undergone the basic training, but who have not yet passed should have to wear the equivalent of L plates (a trainee armband, for example) and be allowed to work only under proper supervision by qualified personnel. Those who fail to pass after a specified period of time – unless there are exceptional circumstances – should lose their licence, as is the case in the Netherlands where the diploma must be awarded within 12 months (Weber, 2002). Central to this system should also be an independent body that oversees the assessment of candidates for licensing.

There are also officers working in shopping centres, stores, accident and emergency departments in hospitals and other venues where they regularly come into contact with the public and where additional skills are required. Such officers often have to make use of their legal tools to arrest and might have to use force more frequently. They are confronted with dangerous situations on a much more regular basis. These are also likely to lead them to court to give evidence. For this type of security officer additional training should be mandated and should include more advanced training in legal skills vis-à-vis arrest and force. It should also cover control and restraint so that officers can adequately and properly restrain a person who is resisting. Courts can be very intimidating places and training should also be given in court skills. Finally for the benefit of the public, security guards should also be qualified in first aid, to the basic workplace level. Those officers working in this type of context should undergo additional training to become Public

Security Officers on top of the GSO level. This would also be achieved through tests of competence. For the supervisor of security officers there should also be additional training in supervisory skills. The higher awards would not necessarily have to be compulsory, rather the career structure in place and other initiatives (see below) could provide incentives for officers to achieve these grades.

One of the defining characteristics of the occupational culture of security officers is inferiority (Button, 2007a). Symbols of rank are very important in the military and the police in denoting status to peers and the public as is the provision of a career structure with incentives to progress. Therefore there should be a designated badge that a person who has completed appropriate training is entitled to wear, and its significance should be reinforced by the introduction of sanctions for wearing a badge to which one is not entitled. This could be on an armband and could also display the officer's licence number (which, like the police displaying their numbers, would greatly aid accountability). For a trainee the armband could indicate their licence number and the letters TSO (Trainee Security Officer) and for qualified officers the number could be accompanied by GSO, PSO, GSOSv, PSOSv or whatever was appropriate. In time these grades would become recognised by the general public. Many purchasers of security would want to know the composition of staff they are likely to get and most importantly the public display would encourage staff to seek advancement to achieve the higher badges.

In Finland they have a form of this type of system, where guards who complete the recognised vocational training are entitled to wear the letter 'A' on their epaulettes (Government Decree on Private Security Services 534/2002 Section 9 (2)). This has proved very successful, encouraging staff to undertake additional training and giving greater status to successful candidates. One must also remember that this is on top of the minimum training for a security guard of 100 hours!

In the previous chapter we noted the lethal weapons, non-lethal weapons, defensive tools and utility tools available to security guards in different contexts. Any use of these should be preceded by additional – sometimes mandatory – training. A security officer who has completed such advanced training should be entitled to wear an approved badge on their armband and it should also be confirmed on their licence. It could be as simple as those who have undergone the appropriate training in the use of handcuffs being entitled to wear a badge with a picture of a handcuff on it. This could ensure that in potentially difficult situations it was clear to members of the public that security officers were competent for the task. Counter-terrorism could be another area in which officers could earn a badge for undertaking an appropriate training course that would enable them to participate in certain counter-terrorist work such as Project Griffin (this covers identifying intelligence on those in likely target areas engaging in suspicious activities, communications, emergency response and helping to secure a cordon in the event of a major incident) (Metropolitan Police, 2007). The badges to denote status and competency will provide a much needed incentive to achieve qualifications in the industry, create a stronger career structure and most importantly make clear to the public what the officer is competent in. To further embed them the regulatory body might also consider mandating or advising minimum standards of pay to further enhance a career structure and progression.

Building a professional infrastructure

Regulation only provides the basic foundations for enhancing the effectiveness of security. To improve it even further a professional infrastructure needs to be created to instigate a culture of continuous improvement. In Chapter 5 we looked at the weaknesses of the professional infrastructure for security management. These included the lack of a coherent body of knowledge, the absence of any recognised professional representation, inadequate training, no proper entrance requirements and an ill-enforced code of governance and ethics. It is clear from this that what needs to be done is very similar to the earlier professionalisation of personnel management, and McGee (2006) drawing upon the work of Larson (1977) illustrates how the 'professional project' of personnel managers led to the emergence of the new profession of Human Resource Management (HRM). This was achieved without any direct statutory intervention, which is the other – regulatory route – model to achieving a profession. For security management both offer valuable insights on professionalisation, but before an agenda is suggested, it is worth sketching the personnel management project as researched by McGee.

A number of strategies were pursued by personnel managers in order to achieve the status of a profession. McGee contrasts the status of personnel managers in the 1970s and early 1980s to the situation today. He paints a picture of personnel managers with few if any specialist qualifications, neglected in strategic decision-making in the organisations they served, criticised by major government reports, such as the Donavon Commission as lacking professionalism, and represented by more than one professional association, with many not represented at all. This contrasts with a situation in which HRM has become a dominant model and where personnel functions are integrated into the broader strategic goals of the organisation. It would be a fair criticism to note that many changes from personnel to HRM have been little more than name changes (Armstrong, 2006). The two main representative associations, the Institute of Personnel Management and the Institute of Training and Development, merged in 1994 to create the Institute of Personnel and Development, which has since achieved the prestigious 'Chartered' status. Almost all those working in HRM/personnel belong to this organisation, which boasts over 130,000 members. The Chartered Institute of Personnel and Development (CIPD) has a staff of 260, lobbies for the profession, is represented on many key forums, produces various publications, and runs seminars and conferences as well as providing a local branch structure.

Most significantly the CIPD has established a membership framework that begins with Affiliate, Associate, Licentiate and Graduate; rising to the Chartered grades from Chartered Member and Chartered Fellow to Chartered Companion. These grades carry weight, with many job advertisements specifying a particular level or an expectation that such a grade will be achieved. The membership grades – depending upon the level – can be achieved through training, higher education and assessment of professional competence. The CIPD also has a code of ethics that if breached can lead to expulsion, something which in many situations means an end to working in HRM/personnel. The CIPD also does much to manage the image of the profession to ensure it is portrayed in an appropriate light.

However, it is not just personnel managers in the field of management support functions who have managed to turn their occupation into a profession. Health and Safety managers are far advanced towards professionalisation, with the Institution of Occupational Safety and Health (IOSH) having 32,000 members worldwide and 12,500 Chartered Safety and Health Practitioners. Membership at this level is based on the completion of a recognised course, usually a degree, masters or Level 4 course combined with two years experience and regular continuous professional development. IOSH achieved Chartered status in 2005.

A route map to a profession

The above offers a 'route map' to a profession for security managers and the broader industry. The first and easiest step is for the bodies that seek to represent security managers to begin to merge. This process has already begun with the merger of the IISec and the SyI in

January 2008. Other bodies operating in the UK security industry in this area should also merge.

The newly merged SyI faces a number of challenges in its first few years if it is to create the foundations for a profession. One of the first tasks will be to create a membership structure based on training, higher education and professional competence. Learning routes will need to be created such that people looking to enter the 'profession' undergo an appropriate training course or higher education award to achieve an entry membership. Given the large numbers already working in security, opportunities should also be created for those who can demonstrate

Table 9.2 A model learning route for security managers

Membership category	Training or academic course	Experience
Associate	Those studying on a recognised training course or university course	None
Member	Completion of Level 3 Security Management Training Course Level 3	Holding relevant security management position
Diplomate	Completion of Level 4 Security Management Training Course Completion of Foundation Degree, Diploma of Higher Education or Certificate of Higher Education Level 4/C/I	At least two years in a security management position or equivalent
Fellow	Completion of Level 5 Security Management Training Course Completion of Honours Degree or Masters Degree in Security Management Level 5/H/M	At least four years in a security management position or equivalent
Companion		Awarded for outstanding achievement in furthering the profession of security management to those who have achieved Fellow level

professional competence through an accreditation process. Table 9.2 illustrates how a membership framework could look. The categories of membership could be achieved by recognised training courses or academic awards at the appropriate level in the NQF. Alternative standards could be created that enable those with appropriate experience to demonstrate their competencies at that level through accredited prior experiential learning. Alongside the training and academic awards for the higher levels there should also be minimum levels of experience. Finally, as with the CIPD a 'Companion' grade should be created that is awarded to those who have reached Fellow and made an outstanding contribution to the progress of the profession.

It is not enough, however, just to create such a framework. The next step is to market it and enforce. All security managers should be encouraged to join and those with the responsibility for recruiting new managers should specify the appropriate level of membership as an essential requirement. As discussed earlier, the Member level should ideally be compulsory through statutory licensing.

The new merged body should learn from other representative associations and offer a range of services that further enhance professionalism. These might include:

- Annual conference
- Seminars on appropriate subjects
- Training
- Branch structure for knowledge transfer/networking
- Accreditation of training and academic courses
- Professional magazine
- Professional journal
- Conduct, commission and disseminate research
- Develop online resources
- Develop best practice and guides to specific security functions
- Sell publications at discount
- Publicise job opportunities
- Provide email alerts on latest information

Elements of this list already exist and deals could be pursued to benefit members. For example, the membership fee could be set such that it included distribution of one of the main security magazines, *Professional Security* or *SMT*, rather than the association producing another journal from scratch. *Security Journal*, an academic publication, could also fit this category. A clear priority will be an annual conference of security professionals which provides opportunities to share knowledge on the latest developments in enhancing security. Such an event should invite members to present papers in panel groups alongside invited speakers covering specific issues in plenary forums. It should build upon the standards of the ASIS annual convention, aspire to the standards of academic conferences such as that of the British Society of Criminology and should be held each year at an easily accessible venue within the UK and at a price that the average security manager could afford. This is a woefully inadequate aspect of the UK private security industry, where there are many conferences held, but they are frequently very expensive, cater only for a few and generally fail to provide opportunities for the submission of papers.

There is another area where such an association could have a very important role to play in enhancing security. Earlier this book advocated the importance of different types of isomorphic learning. There is a reluctance amongst many in the security industry to discuss security and even more so security failure. Yet all security managers can learn from others' experience. There are some bodies that seek to achieve this, such as the Risk and Security Management Forum. However, this is very elitist. A major function of the association should be to develop a branch structure where security managers can openly discuss security issues, particularly security failure, 'behind the wire', amongst other security professionals under so-called 'Chatham House Rules' (what is discussed is not discussed outside the room). The development of such networks will greatly enhance isomorphic learning and should lead to positive improvements of security overall.

These networks should also provide a basis for enhancing the capacity for dealing with terrorism. A network of security managers vetted, trained and in regular communication provides a potentially much more effective link for cooperation between the public law enforcement community and the private security industry. Terrorism is the headline area of cooperation, but there are also many more mundane areas of crime and disorder where such networks will be useful for both sides.

It is important to link such developments to codes of ethics and enforce the 'Chatham House Rules' to the code of conduct. This, however, is just one aspect of what the code should cover. It should also include – among many other aspects – the exercise of functions with honesty and integrity, adherence to appropriate laws and regulations, commitment to abide by the rules of the association, commitments to develop professionally, and respect for the rights of minority groups and the importance of human rights. The new association should set such a code, publicise it to members and actively enforce it.

Most established professions have centres of excellence in some form, which conduct research, identify best practice and have established networks for disseminating that best practice. In the UK security industry there have been some good examples, such as the Perpetuity Group that has conducted an extensive range of research projects, the outputs of which are available on their website and many of which have been utilised for this book. The University of Portsmouth and Leicester University have small trickles of research and at Bucks New University the specialised area of events security has spawned the Centre for Crowd Management and Security. However, what is required is much more investment in research and for more centres of excellence in security to be established in universities. In Australia, a country with a population approximately one-third that of the UK, there are already a number of developments in university departments conducting security related research and acting as centres of excellence. These were initiated with substantial public funding, and include the Australian National University Regulatory Institutions Network, University of Woollongong Centre for Transnational Crime Prevention and Griffith University Centre of Excellence in Policing and Security. The establishment of institutes like these should also be pursued in the UK.

Redefining security management

In Chapter 4 we explored Gill et al.'s (2007) identification of a new breed of security managers they termed the 'new entrepreneurs'. Briggs and Edwards (2006: title page) have also written of the resilience of the security business and observed how, '...the business of security has shifted from protecting companies from risks, to being the new source of competitive advantage'. The extent to which these new approaches to managing security have penetrated managers' mentalities is difficult to determine, and Briggs and Edwards probably overestimate the importance of this approach. Nevertheless, following the CIPD model it would seem timely to develop the 'new entrepreneurs' and some of the practices advocated by Briggs and Edwards into the security management equivalent of HRM – which I shall call Security Risk Management (SRM). There are five principles to redefining security management as Security Risk Management:

1. *Integrating security in the core aims of the organisation.* Security risk management should be aligned as far as practicable in the broader

aims of the organisation rather than acting as a separate function that services the main organisation. This also requires security managers to demonstrate generic business skills, engage in their lexicon and be able to exert influence on other members of management and the board. This also means that security specialists should be represented on the board and in some organisations where security is a particularly important issue, a security director should actually be on the board.

2. *Using security risk management to secure competitive advantage.* The effective use of the most up-to-date security risk management techniques can bring competitive advantage. For example, a 10 per cent reduction in losses to a large retail company currently losing £50 million per annum amounts to a £5 million saving (minus any additional costs of the new technique).

3. *Evidence-based actions.* To achieve the above it is necessary for SRMs to engage more in research and to learn from the experience of peers. They need to be aware of what works, to monitor research on the latest security (and other relevant) strategies, to conduct isomorphic learning and to share experience at appropriate professional events. They also need to be more willing to embrace research in order to assess the effectiveness of their strategies.

4. *Using metrics to monitor performance.* To maximise evidence-based action SRMs also need to maximise the use of metrics to enable performance of different strategies to be monitored and to enable ROI decisions to be made more effectively.

5. *Agents for cultural change.* Underpinning all of the above is the need for a root and branch cultural change to the way security is done. SRMs are key to this change and need to emulate a model of professional practice that breeds greater respect and influence. They have the ability to change the way security is done in their organisations and across society as a whole.

The idea for the creation of MScs, MAs, specialist MBAs, taught MPhils or new MSAs that integrate the latest security and management thinking was floated earlier in this chapter. Central to transforming the practice of security towards SRM is the need for general managers to understand the benefits of an SRM approach. To achieve this, management courses, in particular MBAs, should provide SRM options as part of other management courses. In parallel, security managers need to embrace the latest management thinking by studying courses that expose them not just to the latest in security research and theory, but also to developments in management thinking. In short there need to be fundamental changes in the thinking of many general and security managers, as well as new and amended academic provision.

This section has set out the route map to security management becoming a profession and to the new model of doing security: SRM. Creating more effective security can only partly be achieved by regulation and greater professionalism. Security inequity also needs to be addressed and the final section will explore some of the approaches that could be pursued to achieve this.

Tackling security inequity through security unions

The final part of this chapter is dedicated to some of the most radical ideas on enhancing security in society, which have developed around the subject of addressing security inequity. This builds upon the ideas of the nodal perspective, most clearly set out by Johnston and Shearing (2003) in *Governing Security*. The authors advocate a theoretical framework that opens possibilities for the development of more normative proposals. They are particularly interested in the possibility of concentrating power at a nodal level to steer governance. They raise the prospect of giving capacity to nodes that do not have adequate resources to deliver security, based on 'bottom-up' initiatives where communities are given the resources to act on their own account. Juxtaposed to this, alternative dispute resolution mechanisms are advocated, based upon non-state involvement. The model is influenced a great deal by the experience of the South African town of Zwelethemba as well as more generic corporate models of security management. Essentially what they are advocating as a model for further debate is for communities to establish their own security systems, where they can use their resources to manage security risks and apply sanctions for breach of rules in the same way as would any corporate body.

The theoretical model that Shearing and others have developed has come under most criticism from Loader and Walker (2007), who instead advocate 'state anchored pluralism' in which the state would retain a greater role. Loader and Walker go to great lengths to demonstrate the way in which a role for the state can be created that overcomes some of the traditional criticisms. Nevertheless whatever role the state is given there is still extensive evidence that state-led initiatives come to be dominated by professional interests, rather than the interests of those they serve, frequently vis-à-vis the police (Monkennen, 1981; Nalla, 1992; Hope, 2001; Loveday, 2006; Hughes, 2007; Hughes and

Rowe, 2007; Richardson, 2008). Elected representatives may often be sidelined and themselves may frequently not 'represent' local interests (Hughes and Rowe, 2007; Richardson, 2008). State-led initiatives can also be slow to respond to public demands (if they respond at all). Take the example of the 'Bobby on the beat' in the UK. Since the early 1990s there have been increasing public demands for a greater police presence on the streets. The police have been unable to meet this demand, and the last two decades have seen expanding use of private security patrols, local authority warden schemes and – in extreme cases – vigilantism to fill this gap (Sharp and Wilson, 2000; Crawford et al., 2005). Community Support Officers arrived only in 2003 as a strategy by the police to address this demand – and these are only gradually being extended throughout the UK (Crawford et al., 2005). For various reasons – resources, police working practices, bureaucracy – the police have been unable to meet public demands and there are still many areas where the patrolling presence is inadequate. It is therefore no surprise that some have been tempted by entrepreneurial security patrols and indeed vigilantism, neither of which are generally supported by the police (Noaks, 2000; Sharp and Wilson, 2000; Button, 2002; Johnston, 1996).

It is also worth considering the experience of Crime and Disorder Reduction Partnerships (CDRP) established under the Crime and Disorder Act 1998. These were created to bring the police and local authorities together with other relevant state agencies, non-government organisations and community groups to develop a crime and disorder strategy at a local level (unitary local authority or district council). CDRPs were envisaged to actively engage with the community, giving local residents a real voice and influence in the strategy. The reality has not reflected this community involvement ideal. First, CDRPs have national objectives that they must pursue, whether or not they are a real priority locally (Hughes and Rowe, 2007). Second, the locally identified objectives are often defined by the key agencies – particularly the police – and public involvement in their development is very limited. Indeed, many members of the public are unaware of what a CDRP is let alone how to influence it (Loveday, 2006). There is generally no clear stake for the wider public in the development of a CDRP strategy. There are many positives to come from CDRPs but increasing community engagement in strategies to enhance security is not one of them.

Underpinning much of Loader and Walker's critique of the nodal governance model is a belief that bottom-up initiatives are intrinsically inferior to state-led initiatives. This ignores a growing body of evidence examining the voluntary spirit in some deprived areas and the 'third sector', which has a long history of success with cooperatives, credit unions and other forms of mutualism (see Richardson, 2008; Borzaga and Defourny, 2001; Birchall, 2001, 2003; Westall, 2001; Pearce, 2003).

Recent research on deprived areas across the UK has shown that there are hundreds of small groups of volunteers involved in assemblages of varying size and sophistication committed to working to improve their localities. Evidence of their having a real impact on improving local services and bringing greater capital investment to their communities was also found (Richardson, 2008).

It is cooperatives, however, that provide the best example of bottom-up initiatives and they have a long history of enabling disadvantaged people to escape poverty. In Brighton in 1826 and more famously in Rochdale in 1844 poor people came together to form the first recognised cooperatives through which they bought food in bulk to enable them to sell to their members at more reasonable prices. By the end of the nineteenth century cooperatives and friendly societies had expanded to create much larger consumer cooperatives that also provided banking and insurance services across – among others – the UK, France, Germany and Italy (Birchall, 2003). Such is the extent of the worldwide cooperative movement today that it is estimated that there are 800 million members of cooperatives in 87 countries with over half the world's population made secure (in a broad sense) as a result of cooperative enterprise (International Cooperative Alliance, 2007). Indeed in a study that explores the potential of cooperatives in the fight against poverty, self-organisation by the poor is considered a major precondition (Birchall, 2003).

It is worth looking more closely at credit unions to illustrate the relevance of cooperatives to providing security. Like security, one of the fundamental needs of the population is access to banking services that enable them to deposit cash, pay bills and, most significantly, occasionally to borrow money. According to the World Council of Credit Unions (the world apex body) there are over 37,000 credit unions with over 100 million members in 87 countries (McCarthy et al., 2001). In the UK many of these are in deprived areas where 'loan sharks' would be the only other lending resource. Credit unions are established and run largely by local communities on a democratic basis, with limited help from the state. They provide a very relevant example of how communities – even in the most deprived areas – can come together to build structures to deal with fundamental needs, often in the face of deviant forces seeking to stop them. The deliverance of security could be built upon similar nodal foundations and there are compelling arguments for the benefits of such a model.

Cooperatives worldwide are built on strong values and principles. The former include: self-help, self-responsibility, democracy, equality, equity and solidarity (Cooperatives UK, 2005). The latter include:

- Voluntary and open membership: they are voluntary organisations open to all regardless of race, politics, religion etc.
- Democratic member control: they are member controlled organisations where all have equal votes and the ability to elect directors and influence policy and decision-making.
- Member economic participation: members control the capital of the organisation and allocate surpluses.
- Autonomy and independence: cooperatives are independent, autonomous organisations controlled by their members.
- Education, training and information: they provide education and training for members, directors and staff to enhance the development of the cooperative.
- Cooperation among cooperatives: cooperatives cooperate to strengthen them all.
- Sustainable development of communities: cooperatives work for sustainable development of their communities.

It is worth expanding on some of the advantages of the cooperative model. Primarily they are run by their members for their members and this creates a much stronger stake in the prospects and activities of the organisation. The ability of any member to participate in decision-making gives them an incentive to become involved and to ensure the cooperative is a success. Most importantly it means that decisions and policies are created within a framework of what members want, rather than what organisations outside the community think people want. Their orientation is embedded in the community and it is the community that they serve rather than distant profit-seeking shareholders. Cooperatives can also be as entrepreneurial and innovative as ordinary companies in responding to market demands by creating new services and products. By the values and principles they pursue and the very nature of the organisation they are able to secure volunteers to their cause. Cooperatives have also been able to enshrine ethical approaches in their business models, such as Fair Trade and care for the environment, to a much greater level than traditional companies.

It is also important to note some of the potential problems of the cooperative model, particularly vis-à-vis helping the most deprived in society. There was criticism in the 1970s that the very poorest had not been reached by cooperatives, although the UN report that made this claim has been heavily criticised for misunderstanding the very nature of cooperatives: most fundamentally, cooperatives do not exist to redistribute wealth. Further research highlighted the fact that in some countries political interference challenged the principles of cooperatives, such as voluntary and open membership, inhibiting their impact on the very poorest in society. Nevertheless there were legitimate concerns over reaching the very poorest and these led to the emergence of a new paradigm in the 1980s of some strong cooperatives based on laws that confirm their autonomy and member-owned status, with the UN recognising their importance in a bottom-up approach to address poverty (Birchall, 2003). This highlights the need for external help, active members and quality membership in developing structures targeted at the most deprived.

It is also worth acknowledging at this point that no security co-operatives that fit this model have yet been developed (at least as far as I am aware). A critic might argue that this shows that the idea must be flawed. However, for cooperatives to be successful resources must be available to enable structures and services to be developed. By their very nature communities in deprived areas lacking security, finance and many other goods and services don't have access to such resources. This is why a degree of state intervention in providing funds is vital in order to stimulate such initiatives; without this they will not prosper. Nonetheless, even in deprived areas there is some evidence that communities are willing to contribute small sums of money towards security (Noaks, 2000; Sharp and Wilson, 2000).

Alongside the cooperative approach there has also been a growing recognition of the benefits of localisation (Hines, 2000). The use of local businesses, local resources, local workers and local people has been championed as a buffer to globalisation. The focus of this approach has largely been on goods and services and on national governments providing a framework to re-diversify their local economies. The provision of security is not often raised in these discussions, but the principles also apply. Local people know best what is wrong with their community and some of the things that can be done to address it. Giving those local people a stake in solving problems also contributes to strengthening democracy. Indeed research has shown that strategies tailored directly to local problems have a much greater impact on enhancing neighbourhood security (Innes and Jones, 2006).

Of all the cooperatives, credit unions could provide the best model for implementing nodal governance or what could be called 'security unions'. These would be formed by a local community, whether residents[2] or businesses, who come together and define the boundaries and aims of their 'security union'. These unions would be run on a democratic basis by their members and would be guided by the broader principles and values of the cooperative movement. The membership would elect a board to set the strategic direction of the union. Small unions would be run by the membership and larger unions would be run by staff appointed by the members. A key member of staff would be the security manager who would undertake functions for the membership as in a commercial or public organisation. This would include defining the problem, developing a strategy and monitoring the effectiveness of that strategy. Depending upon the available resources of the security union there might be other staff and contracts with security firms and other relevant organisations to provide services that enhance security. Additionally the cooperative orientation would enable the union to bring in volunteers from the local community to undertake other functions. Some of the services provided by the unions would be free, others would be at a discounted price, while others would be provided at a commercial rate. Strategies that the security union might undertake include:

- Provision of advice and security products (at discount or free) to enhance household security
- Funding of activities for young people to divert from anti-social behaviour
- Public CCTV systems
- Reassurance patrols by security officers
- Reassurance patrols by residents
- Resolution of low level deviance

The greatest challenge would be funding. Security unions would get part of their resources through charging for the services they provide, but as observed above deprived areas would be unable to provide the necessary resources and there would have to be some element of state intervention. In environmental taxation there is a growing emphasis on the polluter pays principle. With security there should be a tax on a sliding scale depending upon the extent of security pollution; the greater the security pollution the higher the tax. Thus a gated community would be taxed at a higher than average rate and some areas might not be taxed at all. If we consider that the security market in the UK alone was worth £6 billion in 2005, if the tax amounted to 2 per cent of that, this would produce a fund of £120 million for redistribution (Security Park, n.d.). This could be added to by the government from other funds.

These monies would then be handled by a specialist body similar to the Big Lottery Fund which distributes some of the money for good causes secured through the National Lottery (Big Lottery Fund, n.d.). This money would be used to provide resources to help run and facilitate security unions, such as for the training of their active members. Most importantly it would be used to provide grants to security unions to facilitate projects. This funding could be used as a carrot and stick to ensure that security unions keep to their central values. It would be directed at the most deprived and needy areas, although it will be important not just to rely on the indices of deprivation – as Stenson (2002) has shown, even relatively better off areas have pockets of deprivation that are not always acknowledged. Indeed, to ensure that security unions are as effective as possible there might be benefits in linking more affluent and less affluent areas to increase the pool of capable and committed volunteers available to run the security unions.

Other measures can also be pursued to further enhance the effectiveness of security unions. There is a growing recognition of corporate social responsibility in contributing towards the fight against crime. This is to do with more than just providing resources; it is also relevant to the way in which companies operate and interact with communities (Hardie and Hobbs, 2002). Tax breaks should be provided for companies that fund projects in security unions that contribute towards greater security. There are already many examples of companies undertaking such socially-responsible behaviour, such as the National Grid which has created a programme to retrain prisoners for its business. The programme has subsequently expanded to 80 businesses in five sectors, retraining over 1000 prisoners, with a resultant re-offending rate of 7 per cent compared to the 75 per cent which is more generally the rule (Smith Institute, 2007).

Organisations should be encouraged through tax breaks to allow their staff with appropriate expertise to offer some of their time to the security unions. There are already some companies, such as KPMG, who encourage their staff to engage in appropriate initiatives as part of their broader duties.

So how would a security union be established and how might it work? After the establishment of a fund and a network to provide expertise in the development of a security union, most likely through existing cooperative structures, there would need to be a marketing campaign to highlight the opportunities and benefits. Existing co-operatives or other community groups might then mobilise communities to establish security unions. Once established they could start operating, bid for funds and – depending upon success in funding and other income-generating activities – employ appropriate staff. Let us assume a security union in a medium-sized city, covering an area of two unitary authority sized electoral wards (about 20,000 electors), which has an annual income of £150,000 drawn from national grants, the local authority, business donations and trading activity. This could fund 0.5 of a security risk manager (who might be shared with another security union) to manage the security union (£20k), four full-time security officers (£80k) (these could be in-house or contracted), security infrastructure for area and residents and running costs (remaining £50k).

The SRM would conduct a risk assessment, engage with members and the community and develop a strategy to deal with the problems identified. They would also have an important role in securing additional funding. The SRM would act as a link and lobbyer of the local CDRP and police. The security union might invest in security products in areas of concern and make these available to members at discounted prices or free in certain circumstances. The security officers employed could be accredited under the Police Reform Act 2002 Community Safety Accreditation Scheme, giving them additional powers and recognition by the police. The officers could provide a reassuring presence through patrols and engagement with the local community. They would work in partnership with other members of the 'extended police family' such as the police, community support officers, wardens and any other local security. They might also be joined on patrols by local volunteers. They would respond to low level security incidents, such as graffiti, vandalism, rowdy behaviour and so on, deal with them and where necessary undertake investigations. Where there are low level disputes, which the police frequently do not have the time to deal with, the security union might provide opportunities to resolve them and impose sanctions. It is also important to reiterate that security unions would not replace existing structures, but would build upon them in the same way as does the security of a shopping centre or a gated community.

These are rough ideas that require a lot more thought and detail. Nevertheless security unions rooted in the principles of cooperation, financed through pump-priming funding via a new fund, could become the vehicle through which security inequity between nodes could be addressed. Cooperatives and other forms of social enterprise have proved highly successful in addressing social inequity; all that is required is the political will to help facilitate such structures. It must also be remembered there are other models of social enterprise in the third sector and there might be some locations where another form of social enterprise is more appropriate to deliver security, although cooperatives, in my view, still provide the best vehicle (Westall, 2001; Pearce, 2003). Another issue is the scope of security unions. In Chapter 1 the broader ideas of human security were introduced. This shows that security encompasses much more than just 'traditional' anti-crime measures. The question then is why do security unions have to be restricted to security in the 'traditional view' (that is, anti-crime and deviance)? Could security unions also embrace education, health or housing issues, to name only a few? The answer is that they clearly could, and some areas might be more inclined to establish a 'community union' which embraces a wider range of services of which the aims of a security union are only a part. So long as the core aims of the approach to enhancing security are pursued it does not matter whether they are part of a much broader range of services or security in the 'narrow sense'. Indeed the pursuit of a diversity of different models should provide for interesting evaluations which could be used to determine their relative success.

Conclusion

This chapter has sought to set out how the foundations of security can be rebuilt to enhance its effectiveness. The overall success of security requires action beyond the nodal level. This chapter has therefore set out the need for more effective regulation of security as this is one of the most significant influences of quality. It has also provided a route map to a profession as there are a wide variety of areas where structures need to be reformed and attitudes changed if a more professional orientation is to be created. In doing so the chapter also outlined a new model of security management, Security Risk Management. Finally the chapter tackled the more contentious issue of security inequity and how the creation of security unions rooted in cooperative principles could provide the vehicle to address this. In the next and final chapter the main principles of 'doing security' are brought together to create a model that integrates both national and nodal level developments. The chapter also explores the important – and often forgotten – issue of tackling social inequity as a means of enhancing security in society as a whole.

Notes

1 The University of Laurea in Finland has a philosophy of integrating security thinking into general management for its undergraduate and postgraduate degrees in security management.
2 There may also be scope in certain areas for particular groups to come together, such as the elderly, gay and lesbians and so on.

Part IV
Concluding Comments

10
Concluding Comments and a Model for Doing Security

This book has sought to identify the weaknesses in the delivery of security and to set out a model for a more effective means of doing security. In doing so it should have stimulated new thinking about security and provided new insights on how security can be best delivered in society. In this final chapter some of the key findings from this book will be revisited, and it will seek to draw together the various reforms advocated throughout Part III to create a model for doing security. It will also set out an agenda of issues that require further research. We will end with a consideration of the final – and increasingly neglected – aspect of doing security, tackling social decay. Before we embark upon this, however, it would be useful to remind ourselves of the theoretical context in which this book has been pursued.

In Part I, we looked at the remarkable growth of the scale and role of private security. It was argued that this has stimulated four types of response. The radical negative perspective views the expansion of private security in a negative light and seeks to promote policies that will frustrate or limit its expansion. Similar policies are also the aim of the second perspective, the conservative negative, although the driver is the defence of interests rather than a social or economic rationale. The third perspective identified was the traditional positive, which is the most commonly held view, and which sees the growth of private security as a positive to be encouraged, but does not consider the full implications of the expansion. The final perspective was the nodal positive, which amongst other issues seeks also to grapple with the inequalities of security provision arising from the growth of the private sector. This nodal perspective was explored in some depth and noted as providing the theoretical context for this book. The context laid out in Chapter 1 provided the basis for the next part of the book which sought to 'undo' security by examining some of the failings in its delivery.

The first chapter in Part II examined security failure and the myth of security. This is a largely neglected area of research in need of much more attention. The chapter used some of the data that does exist to illustrate the extent of security failure in aviation security, materialistic crime against commercial and public organisations, protests and stunts and workplace violence. The chapter then explored the much better researched area of accidents and disasters and from the theories that have sought to explain these, identified how these can also be applied to security failure. Three case studies of security failure were considered in more depth: the Gardner Museum Heist, the attacks on 11 September and the intruder at Prince William's twenty-first birthday party. From this and the other findings from this chapter it was possible to identify common causes of security failure in socio-technical systems that deliver security. These included: design, the number of layers of security, inadequate risk management, corruption, incompetence, recruitment, selection and training of staff, staff culture, management and supervision and the capacity of security to deal with incidents.

Chapter 3 then naturally moved on to considering the perpetrators of security failure, malefactors. Again the lack of research was highlighted, but that which has been published in this area, along with anecdotal accounts of crimes and protests, was used to assess shop theft, burglary, fraud, robbery and protest. Many malefactors are opportunity-driven which opens the possibility of changing their behaviour through manipulating the number of opportunities. The assessment focused upon malefactors' views of security and what – if anything – could be done to prevent them from targeting a particular node. The final part of the chapter noted that not all malefactors are opportunistic and can be deflected. It identified the 'D' malefactors, the determined (organised criminals, terrorists, protesters) who are so set on their deviant intentions they will do what is necessary to circumvent any systems that are in place; the drunk and drugged, whose state of mind means that they do not act rationally; and the deranged, dumb and desperate malefactors, who are also irrational actors. The chapter concluded with a case study of the television programme 'The Heist', which provided an example of how determined malefactors go about their task of planning and committing a crime. One of the key findings of this chapter was that for determined malefactors security represents nothing but a hurdle to be jumped or sidestepped. If there is anything that influences them, it is the risk of getting caught and the ability to escape. For opportunistic malefactors, however, strategies can be pursued to influence behaviour so a particular node is not targeted.

The third chapter in Part II examined the human element in the security system. Chapter 2 had shown that the human element often underlies security failure: either security officers or security managers, who are frequently 'sub-prime' in quality. This chapter examined the occupational culture of security officers identifying traits such as a low commitment to the job, poor levels of training and machismo, which combine to undermine the effectiveness of security officers. Security managers were also considered. Many of them lack education and training and are unable to engage with the business needs of the organisations they work for.

In the final chapter of Part II the foundations of security were assessed. We saw here how some of the structures underpinning the delivery of security contribute towards its ineffectiveness. Among these in the UK is the inadequate regulatory structure (although this also applies to many other countries). The argument focused in particular on limited or absent mandatory levels of training. The chapter then explored some of the traits that indicate a profession and demonstrated how the practice of security management falls short. Finally the issue of security inequity was assessed and it was shown how it undermines initiatives to achieve greater security in society.

The four chapters in Part III set out an agenda and model for doing security. The first chapter set out a holistic model for the development of a security system. This was followed by two chapters using the framework of Lukes's three dimensions of power to set out security strategies that can prevent a node from being targeted, by changing malefactors' behaviour and by refocusing the behaviour of malefactors. This is achieved by enhancing the security system (in particular security officers) so that malefactors are deterred from attempting crimes, and making the security element so effective that when it does have to intervene it is highly capable of doing so. These three chapters are best summarised by the model shown in Figure 10.1.

Chapter 9 outlined strategies that can rebuild the foundations of security. The chapter set out the case for more effective regulation through higher training standards, wider regulation and greater consultation. It showed how an understanding of the industry could be utilised to make higher voluntary standards more desirable. An agenda for enhancing the professional structure of security management was outlined and a new model of Security Risk Management was proposed

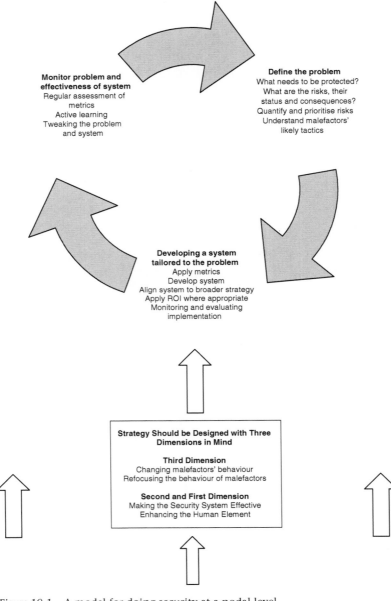

Figure 10.1   A model for doing security at a nodal level

as a means to achieve this. Finally in the most radical proposal of the book, security unions based upon cooperative principles were proposed as a means of tackling security inequity.

A research agenda

Almost all the areas of security touched on in this book would benefit from further research. However, given limited interest amongst practitioners and academics and inevitably limited resources, realism dictates that not all these areas can be researched. Therefore five of the most salient areas requiring further research will be indicated below.

The first area that requires greater consideration is security failure. This book trawled publicly available data and was able to undertake some modelling. However, there needs to be a much more systematic consideration of security failure across a wide range of contexts. Ideally organisations should 'open their books' to researchers to enable a significant number to be studied. This should be able to shed more light on what causes security failure, the most common contributors and what might prevent it from occurring again.

The body of research on malefactors – pioneered by Martin Gill – also needs to be replicated further. The views of different types of malefactors on different types of crimes and other incidents needs to be considered. There is a wealth of data to be gleaned from such research which could be used significantly to enhance security.

Elsewhere I have advocated greater research on security officers (Button, 2007a). This still stands, but security managers are also in need of much more research. We need to ascertain basic data, starting with personal details, career background, education, training, responsibilities, attitudes and so on. As Chapter 4 demonstrated there has only been limited empirical research on security managers and it was carried out over a decade ago. Therefore a large-scale survey of security managers across the private and public sectors should be a priority.

The foundations of security require the development of professional infrastructures. There are many examples of professions that have emerged relatively recently, either largely through statutory intervention (chiropractors) or through the sheer efforts of the profession itself (personnel managers). The research of McGee (2006) provides a marker for further research on other occupations to shed more light on the path to professionalism for security managers.

Finally this book has set out radical proposals for the creation of security unions. This, however, requires more than just research, it also necessitates the resources to establish a pilot security union (or unions) and then detailed research of the impact and success of such a scheme over a longitudinal basis. Such a project requires the commitment and resources of government and the cooperative movement. Now that a research agenda has been set out there is one final issue that can also contribute to greater security.

Tackling social decay

It would not be appropriate to end this book without reference to the neglected issue underlying the cause of much insecurity, the growing social decay in society. I recently attended a meeting of a local interest group in Portsmouth at which the speaker showed a map indicating the highest crime areas. She then showed maps of areas with the highest social security spending, the highest rate of teenage pregnancy, lowest educational achievement, poorest health, lowest income and so on. Most of these maps could be superimposed neatly over one another. In many industrialised countries the pursuit of neo-liberal policies over the last thirty years has culminated in growing social divisions; the creation of large numbers of men and women with no skills or hope of finding meaningful employment, with increasing criminality one of many other associated problems (Reiner, 2007). There are pockets of social decay in most cities in industrialised countries. These have become 'breeding grounds' for crime and have been largely abandoned by mainstream political parties, when they require much stronger political and economic intervention. It would be wrong to argue that a return to a post-war consensus based on Keynesian models of economic delivery will be the saviour of society. Similarly it would be wrong to argue that more effective mechanisms for dong security will succeed alone in turning the tide. There is a need not only to develop the most effective structures of doing security, but also to tackle the underlying causes of the insecurity by improving these pockets of social decay. What those latter requirements are is beyond the scope of this book, but it will certainly require greater state resources, increased intervention and policies that are different to those currently been pursued in the UK. As Reiner (2005) succinctly argued in response to a speech by Sir Ian Blair about policing (but which could equally apply to the kind of security that we want):

> ... [community] cohesion ultimately depends on wider social and cultural processes, including stable employment, the family as a

crucible of responsibility and support, and a sense that the social order is just in its allocation of rewards.

Sir Ian's basic question, 'What kind of police service do we want?' cannot be considered in isolation from the question of the kind of society we have and want. Policing is a symbol, not a source, of the character of a civilisation. Policing cannot be expected to underpin a social order whose foundations have been eroded by Blatcherite neoliberalism.

Bibliography

Adams, J. (1995) *Risk*. London: UCL Press.

ADT Security (n.d.) *Residential*. Retrieved 4 September 2007, from http://www.adt.co.za/residential/residential.html.

Adu-Boakye, K. (2002) 'Private Security and Retail Crime Prevention: an Ethnographic Case Study of Retail Shops in Portsmouth'. MSc dissertation, University of Portsmouth.

Åkerstedt, T. (1991) 'Sleepiness at Work: Effects of Irregular Working', in T.H. Monk (ed.) *Sleep, Sleepiness and Performance*. Chichester: Wiley.

Alfredsson, L., M. Åkerstedt, M. Mattsson and B. Wilborg (1991) 'Self-reported Health and Well Being amongst Night Security Guards: a Comparison with the Working Population'. *Ergonomics*, 34: 525–30.

Andress, M. (2002) *Surviving Security. How to Integrate People, Process and Technology*. Indianapolis: Sams.

APEX (1991) *A European Charter for the Private Security Industry*. London: APEX.

Armstrong, M. (2006) *A Handbook of Human Resource Management Practice*. London: Kogan Page.

Association of Certified Fraud Examiners (2004) *Report to the Nation on Occupational Fraud and Abuse*. Retrieved 15 October 2007, from http://www.acfe.com/documents/2004RttN.pdf.

Association of Chief Police Officers (ACPO) (1995) 'Memorandum of Evidence', in House of Commons Home Affairs Committee (1995) *The Private Security Industry*. Volume II. London: HMSO.

Atkinson, R. (2003) 'Domestication by Cappuccino or a Revenge on Urban Space? Control and Empowerment in the Management of Public Spaces'. *Urban Studies*, 40: 1829–43.

Atkinson, R. and J. Flint (2004) 'The Fortress UK? Gated Communities, the Spatial Revolt of the Elites and Time Space Trajectories of Segregation'. ESRC Centre for Neighbourhood Research Paper 17. Retrieved 3 June 2007, from http://www.bristol.ac.uk/sps/cnrpaperspdf/cnr17pap.pdf.

Atkinson, R., J. Flint, S. Blandy and D. Lister (n.d.) *The Extent and Neighbourhood Impacts of Gated Communities*. Retrieved 20 June 2007, from http://www.bristol.ac.uk/sps/cnrpapersword/report_gated12.doc.

Ayres, I. and J. Braithwaite (1992) *Responsive Regulation*. Oxford: Oxford University Press.

Baldeschwieler, J. (1993) 'Explosive Detection for Commercial Aircraft Security', in P. Wilkinson (ed.) *Technology and Terrorism*. London: Frank Cass.

Bamfield, J. (1994) 'Electronic Article Surveillance: Management Learning in Curbing Theft', in M. Gill (ed.) *Crime at Work: Studies in Security and Crime Prevention*. Leicester: Perpetuity Press.

Bamfield, J. (1998) 'Retail Civil Recovery: Filling a Deficit in the Criminal Justice System?' *International Journal of Risk, Security and Crime Prevention*, 3: 257–67.

Bayley, D.H. (1994) *Police for the Future*. Oxford: Oxford University Press.

Bayley, D.H. and C.D. Shearing (1996) 'The Future of Policing'. *Law and Society Review*, 30: 585–606.

BBC (2004) *Airport Security Lapses Exposed*. Retrieved 20 August 2007, from http://www.bbc.co.uk/insideout/yorkslincs/series5/airport_security_humberside.shtml.

BBC News (2001a) *Robbie Williams Attacked on Stage*. Retrieved 13 August 2007, from http://news.bbc.co.uk/1/hi/entertainment/1184351.stm.

BBC News (2001b) *One in Three 'Lie on CVs'*. Retrieved 18 October 2007, from http://news.bbc.co.uk/1/hi/business/1475221.stm.

BBC News (2002a) *Swedish Airport Security Praised*. Retrieved 20 August 2007, from http://news.bbc.co.uk/1/hi/world/europe/2225741.stm.

BBC News (2002b) *Weapons 'Smuggled onto UK Flight'*. Retrieved 20 August 2007, from http://news.bbc.co.uk/1/hi/uk/2243651.stm.

BBC News (2002c) *FBI Names LA Airport Gunman*. Retrieved 26 October 2007, from http://news.bbc.co.uk/1/hi/world/americas/2096241.stm.

BBC News (2002d) *England Fans Shot by Security Guards*. Retrieved 9 August 2007, from http://news.bbc.co.uk/1/hi/uk/2321961.stm.

BBC News (2003) *Blunkett 'Concerned' by Windsor Stunt*. Retrieved 13 August 2007, from http://news.bbc.co.uk/1/hi/uk/3011856.stm.

BBC News (2004a) *Hopes of Finding Diamond Haul Fade*. Retrieved 22 August 2007, from http://news.bbc.co.uk/1/hi/world/europe/3364911.stm.

BBC News (2004b) *Blair Hit During Commons Protest*. Retrieved 21 August 2007, from http://news.bbc.co.uk/1/hi/uk_politics/3728617.stm.

BBC News (2004c) *Pro-Hunt Protesters Storm Commons*. Retrieved 21 August 2007, from http://news.bbc.co.uk/1/hi/uk_politics/3656524.stm.

BBC News (2004d) *Life Term for Nurse Attacker*. Retrieved 6 November 2007, from http://news.bbc.co.uk/1/hi/scotland/3610823.stm.

BBC News (2005) *Mall Bans Shoppers' Hooded Tops*. Retrieved 3 June 2005, from http://news.bbc.co.uk/1/hi/england/kent/4534903.stm.

BBC News (2006) *Dentist Struck Off for NHS Fraud*. Retrieved 18 October 2007, from http://news.bbc.co.uk/1/hi/england/london/5181706.stm.

BBC News (2007a) *Airport 'Security Lapses' Exposed*. Retrieved 20 August 2007, from http://news.bbc.co.uk/1/hi/england/west_midlands/6716187.stm.

BBC News (2007b) *Nurse Attack Drink Driver Fined*. Retrieved 6 November 2007, from http://news.bbc.co.uk/1/hi/scotland/highlands_and_islands/6398313.stm.

BBC News (2007c) *Jacqui Smith Denies Cover Up*. Retrieved 14 November 2007, from http://news.bbc.co.uk/1/hi/programmes/bbc_parliament/3188796.stm.

Beck, A. (2006) 'Shrinkage and Radio Frequency Identification (RFID): Prospects, Problems and Practicalities', in M. Gill (ed.) (2006) *The Handbook of Security*. Basingstoke: Palgrave.

Beck, A. (2007) 'The Emperor Has No Clothes: What Future Role for Technology in Reducing Retail Shrinkage?' *Security Journal*, 20: 57–61.

Beck, A. and A. Willis (1995) *Crime and Security – Managing the Risk to Safe Shopping*. Leicester: Perpetuity Press.

Becker, T.M. (1974) 'The Place of Private Police in Society: an Area of Research for the Social Sciences'. *Social Problems*, 21: 438–53.

Bennett, T. and R. Wright (1984) *Burglars on Burglary*. Aldershot: Gower.

Bickman, L. (1974) 'The Social Power of a Uniform'. *Journal of Applied Social Psychology*, 4: 47–61.

Big Lottery Fund (n.d.) *About Big*. Retrieved 5 December 2007, from http://www.biglotteryfund.org.uk/index/about.htm.

Birchall, J. (ed.) (2001) *The New Mutualism in Public Policy*. London: Routledge.

Birchall, J. (2003) *Rediscovering the Cooperative Advantage: Poverty Reduction through Self-Help*. Geneva: International Labour Organisation. Retrieved 7 December 2007, from http://www.ilo.org/dyn/empent/docs/F1406653842/poverty%20-%20coops%20birchall%20090103.pdf.

Bjerner, B., A. Holm and A. Swensson (1955) 'Diurnal Variation of Mental Performance: a Study of Three Shift Workers'. *British Journal of Industrial Medicine*, 12: 103–10.

Blakely, E.J. and M.J. Snyder (1997) *Fortress America – Gated Communities in the United States*. Washington DC: Brookings Institution Press.

Borodzicz, E.P. (2005) *Risk, Crisis and Security Management*. Chichester: Wiley.

Borzaga, C. and J. Defourny (eds) (2001) *The Emergence of Social Enterprise*. London: Routledge.

Boston.com News (2005) *The Gardner Heist*. Retrieved 15 August 2007, from http://www.boston.com/news/specials/gardner_heist/heist/.

Bowden, T. (1978) *Beyond the Limits of the Law*. Harmondsworth: Penguin.

Bowers, K. and S. Johnson (2005) 'Using Publicity for Preventative Purposes', in N. Tilley (ed.) *Handbook of Crime Prevention and Community Safety*. Cullompton: Willan.

Brand Failures and Lessons Learned (2006) *Brand PR Failures: Perrier's Benzene Contamination*. Retrieved 14 May 2008, from http://brandfailures.blogspot.com/2006/12/brand-pr-failures-perriers-benzene.html.

Braun, M.A. and D.J. Lee (1971) 'Private Police Forces: Legal Powers and Limitations'. *University of Chicago Law Review*, 38: 555–82.

Briggs, R. and C. Edwards (2006) *The Business of Resilience*. London: Demos.

Bright, J. (1997) *Turning the Tide: Crime, Community and Prevention*. London: Demos.

The British Computer Society Security Group (n.d.) *Who Are We?* Retrieved 5 December 2007, from http://www.bcs-issg.org.uk/about.html.

British Retail Consortium (2005) *BRC Retail Crime Survey: Cost of Crime Up, Violence against Staff Up*. Retrieved 20 August 2007, from http://www.brc.org.uk/details04.asp?id=766&kcat=&kdata=1.

British Retail Consortium (2007) *Retail Crime Survey Key Facts 2005–2006*. Retrieved 20 August 2007, from http://www.brc.org.uk/showdoc04.asp.

British Standards Institution (2006) *BS 7858: 2006 Security Screening of Individuals Employed in a Security Environment – Code of Practice*. London: BSI.

Browne, R.C. (1949) 'The Day and Night Performance of Teleprinter and Switchboard Operators'. *Journal of Occupational Psychology*, 23: 121–6.

Broughton, F. (1995) 'Oral Evidence to Home Affairs Committee', in Home Affairs Committee. *The Private Security Industry*, Volume II. London: HMSO.

Bunyan, T. (1976) *The Political Police in Britain*. London: Julian Friedmann Publishers.

Burris, S. (2004) 'Governance, Microgovernance and Health'. *Temple Law Revies*, 77: 335–59.

Bushman, B.J. (1984) 'Perceived Symbols of Authority and their Influence on Compliance'. *Journal of Applied Social Psychology*, 14: 501–8.

Business Week Online (2003) *El Al's Security vs the US Approach*. Retrieved 26 October 2007, from http://www.businessweek.com/bwdaily/dnflash/aug2003/nf20030825_5134_db039.htm.

Butler, G. (1994) 'Shoplifters' Views on Security: Lessons for Crime Prevention', in M. Gill (ed.) *Crime at Work*. Leicester: Perpetuity Press.

Button, M. (2002) *Private Policing*. Cullompton: Willan Publishing.

Button, M. (2003) 'Private Security Industry Law in Europe: the Case of Great Britain', in S. Outer and R. Stober (eds) *Recht des Sicherheitsgewerbes*. Köln: Heymanns.

Button, M. (2004) ' "Softly, Softly", Private Security and the Policing of Corporate Space', in R. Hopkins-Burke (ed.) *Hard Cop, Soft Cop: Dilemmas and Debates in Contemporary Policing*. Cullompton: Willan Publishing.

Button, M (2007a) *Security Officers and Policing: Powers, Culture and Control in the Governance of Private Space*. Aldershot: Ashgate.

Button, M. (2007b) 'Assessing the Regulation of Private Security across Europe'. *European Journal of Criminology*, 4: 109–28.

Button, M. and B. George (1994) 'Why Some Organisations Prefer In-House to Contract Security Staff', in M. Gill (ed.) *Crime at Work: Studies in Security and Crime Prevention*. Leicester: Perpetuity Press.

Button, M. and B. George (1998) 'Why Some Organisations Prefer Contract to In-House Security Staff', in M. Gill (ed.) *Crime at Work: Increasing the Risk for Offenders*. Leicester: Perpetuity Press.

Button, M. and B. George (2001) 'Government Regulation in the United Kingdom Private Security Industry: the Myth of Non-Regulation'. *Security Journal*, 14: 55–66.

Button, M. and B. George (2006) 'Regulation of Security: New Models for Analysis'. In M. Gill (ed.) *Handbook of Security*. Basingstoke: Palgrave.

Button, M. and T. John (2002) '"Plural Policing" in Action: a Review of the Policing of Environmental Protests in England and Wales'. *Policing and Society*, 12: 111–21.

Button, M. and H. Park (forthcoming) 'Security Officers and the Policing of Private Space in South Korea: Profile, Powers and Occupational Hazards'. *Policing and Society*.

Button, M., T. John and N. Brearley (2002) 'New Challenges in Public Order Policing: the Professionalisation of Environmental Protest and the Emergence of the Militant Environmental Activist'. *International Journal of the Sociology of the Law*, 30: 17–32.

Button, M., H. Park and J. Lee (2006) 'The Private Security Industry in South Korea: a Familiar Tale of Growth, Gaps and the Need for Better Regulation'. *Security Journal*, 19: 167–79.

Cabinet Office (2002) *Identity Fraud: a Study*. Retrieved 16 October 2007, from http://www.identitycards.gov.uk/downloads/id_fraud-report.pdf.

Cambridge Online Dictionary (n.d.) *Malefactor*. Retrieved 13 January 2008, from http://dictionary.cambridge.org/define.asp?key=48372&dict=CALD.

Carratu (n.d.) *Employee Screening Services*. Retrieved 16 October 2007, from http://www.carratu.com/corporate/employee_vetting.htm#.

Challinger, D. (2006) 'Corporate Security: a Cost or Contributor to the Bottom Line', in M. Gill (ed.) *The Handbook of Security*. Basingstoke: Palgrave.

Chaundhary, V. (2000) 'Any Bombs, Knives, Pepsi? Security Gets Tough at the Olympics'. *Guardian*, 18 September, p. 7.

Chu, Y.K. (1996) 'Triad Societies and the Business Community in Hong Kong'. *International Journal of Risk, Security and Crime Prevention*, 1: 33–40.

City of London Police (2003) *Summary of Final Report – Suitable for Publication. Investigation into the Breach of Security During the 21st Birthday Party for HRH Prince William on Saturday 21st June, 2003*. Retrieved 13 August 2007, from http://www.met.police.uk/reports/Armstrong_Report.pdf.

Clarke, R.V.G. (1980) 'Situational Crime Prevention: Theory and Practice'. *British Journal of Criminology*, 20: 136–47.

Clarke, R.V.G. (ed.) (1992) *Situational Crime Prevention: Successful Case Studies*. New York: Harrow and Heston.

Clarke, R.V.G. (2005) 'Seven Misconceptions of Situational Crime Prevention', in N. Tilley (ed.) *Handbook of Crime Prevention and Community Safety*. Cullompton: Willan.

Clarke, R.V. and P.M. Mayhew (1988) 'The British Gas Suicide Story and its Criminological Implications', in N. Morris and M. Tonry (eds) *Crime and Justice: an Annual Review of Research*, Volume 10. Chicago: University of Chicago Press.

CNN.com (2002) *The Isabella Gardner Museum Heist*. Retrieved 15 August 2007, from http://archives.cnn.com/2002/LAW/11/26/ctv.traces.museum.heist/ index. html.

CoESS (Confederation of European Security Services) (2004) *Annual Report*. Retrieved 20 October 2007, from http://www.coess.org/documents/annual_ report_2004.pdf.

CoESS (2006) *News*. Retrieved 20 October 2007, from http://www.coess.org/.

CoESS/ UNI Europa (2004) *Panoramic Overview of the Private Security Industry in the 25 Member States of the European Union*. Retrieved 20 October 2007, from http://www.coess.org/pdf/panorama1.pdf.

Cohen, A.K. (1955) *Delinquent Boys*. New York: Free Press.

Cohen, L.E. and M. Felson (1979) 'Social Changes and Crime Rates: a Routine Activities Approach'. *American Sociological Review*, 44: 588–608.

Cohen, S. (1985) *Visions of Social Control*. Cambridge: Polity Press.

Colling, R.L. (2001) *Hospital and Healthcare Security*. London: Butterworth-Heinemann.

Commission on Human Security (2003) *Human Security Now*. New York: Commission on Human Security.

Co-operatives UK (2005) *Co-operative Values and Principles*. Retrieved 7 December 2007, from http://www.cooperatives-uk.coop/About/valuesAndPrinciples.

Conklin, J.E. and E. Bittner (1973) 'Burglary in a Suburb'. *Criminology*, 11: 206–32.

Cooke, C.A. (2005) 'Issues Concerning Visibility and Reassurance Provided by the New "Policing" Family'. *Journal of Community and Applied Social Psychology*, 15: 229–40.

Couch, S.R. (1987) 'Selling and Reclaiming State Sovereignty: the Case of Coal and Iron Police'. *The Insurgent Sociologist*, 4: 85–91.

CourtTV.com (u.d.) *Heist #4. World's Biggest Diamond Heist*. Retrieved 22 August 2007, from http://www.courttv.com/onair/shows/impossible_heists/real-heists4.html.

Cox, S. and N. Tait (1991) *Reliability, Safety and Risk Management: an Integrated Approach*. London: Butterworth-Heinemann.

Crawford, A. (1997) *The Local Governance of Crime*. Oxford: Clarendon Press.

Crawford, A., S. Lister, S. Blackburn and J. Burnett (2005) *Plural Policing*. Bristol: The Policy Press.

Cumming, J. and S. Winyard (1984) *Working Insecurity*. Liverpool: Low Pay Unit.

Cunningham, W.C. and T. Taylor (1985) *Private Security and Police in America*. Portland: Chancellor Press.

Cunningham, W.C., J.J. Strauchs and C.W. Van Meter (1990) *Private Security Trends 1970–2000*. Hallcrest Report II. Stoneham: Butterworth-Heinemann.

Daily Mirror (1995) *Scandal of the Job that Pays Nothing*. 20 March, p. 11.

Davis, M. (1990) *City of Quartz*. London: Vintage.

De Waard, J. (1993) 'The Private Security Sector in Fifteen European Countries: Size, Rules and Legislation'. *Security Journal*, 4: 58–62.

De Waard, J. (1999) 'The Private Security Industry in International Perspective'. *European Journal of Criminal Policy and Research*, 7: 143–74.

De Waard, J. and J. van de Hoek (1991) *Private Security Size and Legislation in the Netherlands and Europe*. The Hague: Dutch Ministry of Justice.

Demonbynes, G. and B. Ozler (2002) *Crime and Local Inequality in South Africa. World Bank Policy Research Working Paper 2925*. Retrieved 8 January 2008, from http://www-wds.worldbank.org/servlet/WDSContentServer/WDSP/IB/2002/12/14/000094946_02120504013794/Rendered/PDF/multi0page.pdf.

Department of Health (2003) *A Professional Approach to Managing Security in the NHS*. London: DoH.

Dickins, R., S. Machin and A. Manning (1994) 'Minimum Wages and Employment: a Theoretical Framework with an Application to the UK Wages Councils'. *International Journal of Manpower*, 15: 26–48.

Dickinson, D. (2002) 'Our Mission? To Support the Police', *Security Direct*. 2002–2003: 2.

Dickinson, D. (2003) 'Promised – Not Predicted'. *The 208 Newsletter*, Summer: 3.

Dixon, D. (1997) *Law in Policing Legal Regulation and Police Practices*. Oxford: Clarendon Press.

Doig, A. (2007) *Fraud*. Cullompton: Willan.

Draper, H. (1978) *Private Police*. Sussex: Harvester Press.

Duffin, M., G. Keats and M. Gill (2006) *Identity Theft in the UK: the Offender and Victim Perspective*. Leicester: Perpetuity Research and Consultancy International.

Eco, U. (1976) *A Theory of Semiotics*. Bloomington: Indianapolis University Press.

Ekblom, P. (1999) 'Can We Make Crime Prevention Adaptive by Learning from Other Evolutionary Struggles?' *Studies on Crime and Crime Prevention*, 8: 27–51.

Ekblom, P. (2005) 'Designing Products against Crime', in N. Tilley (ed.) *Handbook of Crime Prevention and Community Safety*. Cullompton: Willan.

Farrell, G. and K. Pease (2006) 'Criminology and Security', in M. Gill (ed.) *The Handbook of Security*. Basingstoke: Palgrave.

Fischer, R.J. and G. Green (1998) *Introduction to Security*. Sixth Edition. Boston, MA: Butterworth Heinemann.

Fletcher, D. (2002) 'Going Private? It's Plain Common Sense', *Security Direct*. 2002–2003: 65.

Fordham, P. (1968) *The Robbers' Tale*. Harmondsworth: Penguin.

Foucault, M. (1977) *Discipline and Punish*. Harmondsworth: Penguin.

Frosdick, S. (2005) 'Basis for Security'. *Stadium and Arena Management*, June.

Frosdick, S. (2007) *Introduction to the Management of Risk*. Portsmouth: Institute of Criminal Justice Studies, University of Portsmouth.

Garcia, M.L. (2006) 'Risk Management', in M. Gill (ed.) *The Handbook of Security*. Basingstoke: Palgrave.

George, B. and M. Button (1997) 'Private Security Regulation – Lessons from Abroad for the United Kingdom'. *International Journal of Risk Security and Crime Prevention*, 2: 109–21.

George, B. and M. Button (2000) *Private Security*. Leicester: Perpetuity Press.

George, B., M. Button and N. Whatford (2003) 'The Impact of September 11th on the UK Business Community'. *Crime Prevention and Community Safety*, 5: 49–60.

Gill, M. (1996) 'Risk, Security and Crime Prevention: an International Forum for Developing Theory and Practice'. *International Journal of Risk, Security and Crime Prevention*, 1: 11–17.

Gill, M. (2000) *Commercial Robbery*. London: Blackstone Press.

Gill, M. (2005a) 'Reducing Capacity to Offend: Restricting Resources for Offending', in N. Tilley (ed.) *Handbook of Crime Prevention and Community Safety*. Cullompton: Willan.

Gill, M. (2005b) *Learning from Fraudsters*. Leicester: Perpetuity Research and Consultancy International.

Gill, M. (ed.) (2006) *The Handbook of Security*. Basingstoke: Palgrave.

Gill, M. (2007) *Shoplifters on Shop Theft: Implications for Retailers*. Leicester: Perpetuity Research and Consultancy International.

Gill, M. and M. Hemming (2004) *Evaluation of CCTV in London Borough of Lewisham*. Leicester: Perpetuity Group.

Gill, M. and R. Mathews (1994) 'Robbers on Robbery: Offenders' Perspectives', in M. Gill (ed.) *Crime at Work: Studies in Security and Crime Prevention*. Leicester: Perpetuity Press.

Gill, M and A. Spriggs (2005) *Assessing the Impact of CCTV*. Home Office Research study 292. Retrieved 5 May 2008, from http://www.homeoffice.gov.uk/rds/pdfs05/hors292.pdf.

Gill, M., T. Burns-Howell, G. Keats and E. Taylor (2007) *Demonstrating the Value of Security*. Leicester: Perpetuity Research and Consultancy International.

Gilling, D. (1997) *Crime Prevention*. London: Routledge.

Gimenez-Salinas, A. (2004) 'New Approaches Regarding Private/Public Security'. *Policing and Society*, 14: 158–74.

Gotbaum, B. (2005) *Undertrained, Underpaid, and Unprepared: Security Officers Report Deficient Safety Standards in Manhattan Office Buildings*. A Report by the Public Advocate of the City of New York, New York: Public Advocate for the City of New York.

Gray, K. (1994) 'Equitable Property'. *Current Legal Problems*, 47: 157–214.

Gray, K. and S.F. Gray (1999a) *Land Law*. London: Butterworth.

Gray, K. and S.F. Gray (1999b) 'Private Property and Public Propriety', in J. McLean (ed.) (1999) *Property and Constitution*. Oxford: Hart Publishing.

Hainmüller, J. and J.M. Lemnitzer (2003) 'Why do Europeans Fly Safer? The Politics of Airport Security in Europe and the US'. *Terrorism and Political Violence*, 15: 1–36.

Hakala, J. (2007) *The Regulation of Manned Private Security: a Transnational Survey of Structure and Focus*. Retrieved 4 March 2008, from http://www.coess.org/pdf/article_on_regulation_survey.pdf.

Hakim, S. and A.J. Buck (1991) *Deterrence of Suburban Burglaries*. Rockville: National Institute of Justice.

Handford, M. (1994) 'Electronic Tagging in Action: a Case Study in Retailing', in M. Gill (ed.) *Crime at Work: Studies in Security and Crime Prevention*. Leicester: Perpetuity Press.

Haney, C., C. Banks and P. Zimbardo (1973) 'Interpersonal Dynamics in a Simulated Prison'. *International Journal of Criminology and Penology*, 1: 69–97.

Hardie, J. and Hobbs, B. (2002) *Partners against Crime: the Role of the Corporate Sector in Tackling Crime*. London: IPPR. Retrieved 7 December 2007, from http://www.popcenter.org/Library/CrimePrevention/Volume%2018/03-Hardie%20&%20Hobbs-Partners%20against%20crime_%20The%20role %20of%20the%20co.pdf.

Harding, R.W. (1997) *Private Prisons and Public Accountability*. Buckingham: Open University Press.

Harma, M.I., J. Ilmarinen, P. Knauth, J. Rutenfranz and O. Hanninen (1986) 'The Effect of Physical Intervention on Adaptation to Shiftwork'. *International Journal of Chronobiology*, 4: 77–110.

Harris, M. (1974) 'Mediators between Frustration and Aggression in a Field Experiment'. *Journal of Experimental Psychology*, 10: 561–71.

Harvey, D. (1990) *The Condition of Post-Modernity*. Oxford: Blackwell.

Hearnden, K. (1993) *The Management of Security in the UK*. Loughborough: Centre for Extension Studies, University of Loughborough and SITO.

Hearnden, K. (1995) 'Multi-tasking in British Business: a Comparative Study of Security and Safety Managers'. *Security Journal*, 6: 123–32.

Helminger, P. (2002) *The Economic Consequences of 11 September and the Economic Dimension of Terrorism*. Brussels: NATO Parliamentary Assembly.

Hemmens, C., J. Maahs, K.E. Scarborough and P.A. Collins (2001) 'Watching the Watchmen: State Regulation of Private Security 1982–1998'. *Security Journal*, 14: 17–28.

Hildebrandt, G., W. Rohmert and J. Rutenfranz (1974) '12 and 24 Hours in Rhythms in Error Frequency of Locomotive Drivers and the Influence of Tierdness'. *International Journal of Chronobiology*, 2: 175–80.

Hines, C. (2000) *Localisation: a Global Manifesto*. London: Earthscan.

Hobbes, T. (1651/1985) *Leviathan*. Harmondsworth: Penguin.

Hobbs, D., P. Hadfield, S. Lister and S. Winlow (2003) *Bouncers: Violence and Governance in the Night-time Economy*. Oxford: Oxford University Press.

Hobsbawm, E., J. (1959) *Primitive Rebels: Studies in Archaic Forms of Social Movement in the 19th and 20th Centuries*. Manchester: Manchester University Press.

Hogg, A., J. McDougall and J. Morgan (1988) *Bullion. Brinks Mat: the Story of Britain's Biggest Gold Robbery*. London: Penguin.

Hollinger, R.C. and J.L. Davis (2006) 'Employee Theft and Staff Dishonesty', in M. Gill (ed.) *The Handbook of Security*. Basingstoke: Palgrave.

Home Office (2000) *The Economic and Social Costs of Crime*. Home Office Research Study 217. Retrieved 21 August 2007, from http://www.homeoffice. gov.uk/rds/pdfs/hors217.pdf.

Hope, T. (2001) 'Community Crime Prevention in Britain: a Strategic Overview'. *Criminal Justice*, 1: 421–39.

House of Commons Defence Committee (1990) *The Physical Security of Military Installations in the United Kingdom*. HC 171. London: HMSO.

House of Commons Defence Committee (2001) *The Threat from Terrorism*. HC 348-I. Volume I. London: The Stationery Office.

House of Commons Home Affairs Committee (1995), *The Private Security Industry*, Vols I and II, HC 17 I and II. London: HMSO.

Huggins, M.K. (2000) 'Urban Violence and Police Privatisation in Brazil: Blended Invisibility'. *Social Justice*, 27: 113–34.

Hughes, G. (2007) *The Politics of Crime and Community*. Basingstoke: Palgrave.

Hughes, G. and M. Rowe (2007) 'Neighbourhood Policing and Community Safety: Researching the Instabilities of the Local Governance of Crime, Disorder and Security in Contemporary UK'. *Criminology and Criminal Justice*, 7: 317–46.

Hyde, D. (2003) 'The Role of "Government" in Regulating, Auditing and Facilitating Private Policing in Late Modernity: the Canadian Experience'. Paper presented to In Search of Security Conference, Montreal, Quebec, February.

Info4security (2007) *Homeowners Warned Over Lock Bumping*. Retrieved 8 November 2007, from http://www.info4security.com/story.asp?storycode= 4115647&enc Code=375599.

Initial Security (n.d) *Security Officers Handbook*.

Innes, M. and N. Fielding (2002) 'From Community to Communicative Policing: "Signal Crimes" and the Problem of Public Reassurance'. *Sociological Research Online*, 7, http://www.socresonline.org.uk/7/2/innes.html.

Innes, M. and V. Jones (2006) *Neighbourhood Strategy and Social Change*. London: Joseph Rowntree Foundation.

Intelligarde the Law Enforcement Company (2007) *Sign Up Now*. Retrieved 10 August 2007, from http://www.intelligarde.org/home.html.

International Alert (2005) *SALW and Private Security Companies in South Eastern Europe: a Cause or Effect of Insecurity*. Retrieved 24 July 2007, from http://www. seesac.org/reports/psc.pdf.

International Co-operative Alliance (2007) *Statistical Information on the Co-operative Movement*. Retrieved 7 December 2007, from http://www.ica.coop/ members/member-stats.html.

Jacobs, J. (1961) *The Death and Life of Great American Cities: the Failure of Town Planning*. London: Cape.

Jason-Lloyd, L. (2003) *Quasi-Policing*. London: Cavendish Publishing.

Johnston, L. (1992) *The Rebirth of Private Policing*. London: Routledge.

Johnston, L. (1996) 'What is Vigilantism?' *British Journal of Criminology*, 36: 220–36.

Johnston, L. (2000) *Policing Britain: Risk, Security and Governance*. London: Longman.

Johnston, L. (2006) 'Transnational Security Governance', in J. Wood and B. Dupont (eds) *Democracy, Society and the Governance of Security*. Cambridge: Cambridge University Press.

Johnston, L. and C.D. Shearing (2003) *Governing Security*. London: Routledge.

Jones, T. and T. Newburn (1998) *Private Security and Public Policing*. Oxford: Clarendon Press.

Kakalik, J. and S. Wildhorn (1971a) *Private Police in the United States, Findings and Recommendations. Volume 1*. Washington DC: Government Printing Office.

Kakalik, J. and S. Wildhorn (1971b) *The Private Police Industry: its Nature and Extent. Volume 2*. Washington DC: Government Printing Office.

Kakalik, J. and S. Wildhorn (1971c) *Current Regulation of Private Police: Regulatory Agency Experience and Views. Volume 3*. Washington DC: Government Printing Office.

Kakalik, J. and S. Wildhorn (1971d) *The Law and Private Police. Volume 4*. Washington DC: Government Printing Office.

Kakalik, J. and S. Wildhorn (1971e) *Special Purpose Public Police. Volume 5*. Washington DC: Government Printing Office.

Kapardis, A. and M. Krambia-Kapardis (2004) 'Enhancing Fraud Prevention and Detection by Profiling Fraud'. *Criminal Behaviour and Mental Health*, 14: 189–201.

Kempa, M. and A. Singh (2008) 'Private Security, Political Economy and the Policing of Race'. *Theoretical Criminology*, 12: 333–54.

Kempa, M., R. Carrier, J. Wood and C.D. Shearing (1999) 'Reflections on the Evolving Concept of "Private Policing"'. *European Journal of Criminal Policy and Research*, 7: 197–223.

Kempa, M., P.C. Stenning and J. Wood (2004) 'Policing Communal Spaces'. *British Journal of Criminology*, 44: 563–81.

Klein, D.E., H. Bruner and H. Holtman (1970) 'Circadian Rhythm of Pilots' Efficiency, and Effects of Multiple Time Zone Travel'. *Aerospace Medicine*, 41: 125–32.

Kontenko, I. (n.d.) *Agent Based Modelling and Simulation of Cyberwarfare between Malefactors and Security Agents in Internet*. Retrieved 3 September 2007, from http://www.comp.glam.ac.uk/ASMTA2005/Proc/pdf/abs-03.pdf.

KPMG Forensic (2007) *Profile of a Fraudster Survey 2007*. Retrieved 12 September 2007, from http://www.kpmg.co.uk/services/f/pubs.cfm#.

Kuratko, D.F., J.S. Hornsby, D.W. Naffziger and R.M. Hodgetts (2000) 'Crime and Small Business: an Exploratory Study of Cost and Prevention Issues in US Firms'. *Journal of Small Business and Management*, 38: 1–13.

Lane, K. (2001) 'Human Resources: is it a Weak Link in the Security Chain of Your Company'. *Security Journal*, 14: 7–16.

Larson, M.S. (1977) *The Rise of Professionalism: a Sociological Analysis*. Berkeley: University of California Press.

Laycock, G. (1991) 'Operation Identification or the Power of Publicity'. *Security Journal*, 2: 67–71.

Learmont, J. (1995) *Review of Prison Service Security in England and Wales and the Escape from Parkhurst Prison on Tuesday 3rd January 1995*. Cm 3020. London: HMSO.

Lee, J. (2006) 'Burglar Decision Making and Target Selection: an Assessment of Residential Vulnerability to Burglary, in the Korean Context'. PhD Thesis, University of Portsmouth.

Lees, L. (1997) 'Agegraphia, Heterotopia, and Vancouver's New Public Library'. *Environment and Planning D: Society and Space*, 15: 321–47.

Leeson, L. (1996) *Rogue Trader*. London: Little, Brown and Co.

Levi, M., J. Burrows, H. Fleming and M. Hopkins (2007) *The Nature, Extent and Economic Impact of Fraud in the UK*. London: ACPO.

Licu, E. and B.S. Fisher (2006) 'The Extent, Nature and Responses to Workplace Violence Globally: Issues and Findings', in M. Gill (ed.) *The Handbook of Security*. Basingstoke: Palgrave.

Lister, S., P. Hadfield, D. Hobbs and S. Winlow (2001) 'Accounting for Bouncers: Occupational Licensing as a Mechanism for Regulation'. *Criminal Justice*, 1: 363–84.

LiveLeak (2007) *Robbie Williams Pushed off Stage in Germany (full)*. Retrieved 13 August 2007, from http://www.liveleak.com/view?i=f00_1181285466.

Loader, I. (1996) 'Private Security and the Demand for Protection in Contemporary Britain'. *Policing and Society*, 7: 143–62.

Loader, I. (1997) 'Policing and the Social: Questions of Symbolic Power'. *British Journal of Sociology*, 48: 1–18.

Loader, I. (1999) 'Consumer Culture and the Commodification of Policing and Security'. *Sociology*, 33: 372–93.

Loader, I. and N. Walker (2007) *Civilizing Security*. Cambridge: Cambridge University Press.

Loveday, B. (1991) 'Police and Government in the 1990s'. *Social Policy and Administration*, 25: 4–11.

Loveday, B. (2006) 'Learning from the 2004 Crime Audit. An Evaluation of the National Community Safety Plan 2006–2008 and Current Impediments to the Effective Delivery of Community Safety Strategy by Local Crime Reduction Partnerships'. *Crime Prevention and Community Safety*, 8: 188–201.

Lukes, S. (1974) *Power: a Radical View*. London: Macmillan.

Macauley, S. (1986) 'Private Government', in L. Lipson and S. Wheeler (eds) *Law and the Social Sciences*. New York: Russell Sage Foundation.

Macpherson, W. (1999) *The Stephen Lawrence Enquiry*. Retrieved 5 December 2007, from http://www.archive.official-documents.co.uk/document/cm42/4262/4262.htm.

Maguire, M. and T. Bennett (1982) *Burglary in a Dwelling*. London: Heinemann.

Manunta, G. (1996) 'The Case Against: Security Management is Not a Profession'. *International Journal of Risk, Security and Crime Prevention*, 1: 233–40.

Manunta, G. (1999) 'What is Security?' *Security Journal*, 12: 57–66.

Mauro, R. (1984) 'The Constable's New Clothes: Effects of Uniforms on Perceptions and Problems of Police Officers'. *Journal of Applied Social Psychology*, 14: 42–56.

Massey, J. (2005), 'Patrols and Control in New Urban Space'. Paper presented to the British Society of Criminology Conference, Leeds, 14 July.

Mayhew, P., R.V.G. Clarke, A. Sturman and J.M. Hough (1976) *Crime as Opportunity*. London: HMSO.

McCarthy, O., R. Briscoe and M. Ward (2001) 'Mutuality through Credit Unions', in J. Birchall (ed.) *The New Mutualism in Public Policy*. London: Routledge.

McGee, A. (2006) 'Corporate Security's Professional Project: an Examination of the Modern Condition of Corporate Security Management, and the Potential for Further Professionalisation of the Occupation'. MSc Thesis, Cranfield University.

McLeod, R. (2002), *Parapolice – a Revolution in the Business of Law Enforcement*. Toronto: Boheme Press.

McDonald, A.D., J.C. McDonald, B. Armstrong, N.M. Cherry, R. Cote and J. Lavoie (1988) 'Foetal Death and Work in Pregnancy'. *British Journal of Industrial Medicine*, 45: 148–57.

Metropolitan Police (2007) *Project Griffin*. Retrieved 8 January 2008, from http://www.met.police.uk/westminster/project_griffin.htm.

Michael, D. (2002) 'A Sense of Security? The Ideology and Accountability of Private Security Officers'. PhD Thesis, London School of Economics.

Micucci, A. (1998) 'A Typology of Private Policing Operational Styles'. *Journal of Criminal Justice*, 26: 41–51.

Milgram, S. (1975) *Obedience to Authority*. New York: Harper Torchbooks.

Millerson, G. (1964) *The Qualifying Associations: a Study in Professionalisation*. London: Routledge and Kegan Paul.

Miyazawa, S. (1991) 'The Private Sector and Law Enforcement in Japan', in W.T. Gormley (ed.) *Privatisation and its Alternative*. Madison: University of Wisconsin Press.

Monk, T.H. (ed.) (1991) *Sleep, Sleepiness and Performance*. Chichester: Wiley.

Monkennen, E.H. (1981) *The Police in Urban America 1860–1920*. Cambridge: Cambridge University Press.

Murray, G. (1993) *Enemies of the State*. London: Simon and Schuster.

Murphy, E. and R. Dingwall (2007) 'The Ethics of Ethnography', in P. Atkinson, A. Coffey, S. Delamont, J. Lofland and L. Lofland (eds) *Handbook of Ethnography*. London: Sage.

Nalla, M.K. (1992) 'Perspectives on the Growth of Police Bureaucracies 1948–1983: an Examination of Three Explanations'. *Policing and Society*, 3: 51–61.

Nalla, M. and M. Morash (2002) 'Assessing the Scope of Corporate Security: Common Practices and Relationships with other Business Functions'. *Security Journal*, 15: 7–19.

National Advisory Committee on Criminal Justice Standards and Goals (1976) *Private Security. Report of the Task Force on Private Security*. Washington DC: Government Printing Office.

National Audit Office (2003) *A Safer Place to Work: Protecting NHS Hospital and Ambulance Staff from Aggression*, London: The Stationery Office. Retrieved 20 February 2007, from http://www.nao.org.uk/publications/nao_reports/02-03/0203527.pdf.

National Commission on the Terrorist Attacks upon the United States (2004) *The 9/11 Commission Report*. New York: WM Norton.

National Retail Federation (2007) *2007 Organized Retail Crime Survey Results*. Washington DC: National Retail Federation.

National Statistics (2007) *Earnings*. Retrieved 10 October 2007, from http://www.statistics.gov.uk/cci/nugget.asp?id=285.

NHSCFSMS (National Heath Service Counter Fraud and Security Management Service) (2005) *Countering Fraud in the NHS: Protecting Resources for Patients*. London: NHSCFSMS.

NHSCFSMS (2006) *Violence against Staff Figures (Per 1000) (05–06)*. Retrieved 6 November 2007, from http://www.cfsms.nhs.uk/doc/sms.general/2005-06_violence_against_NHS_staff_per1000.pdf.

NHSCFSMS (2007) *Countering Fraud in the NHS: Protecting Resources for Patients. 1999–2006 Performance Statistics*. London: CFSMS.

Nee, C. and A. Meenaghan (2006) 'Expert Decision Making in Burglars'. *British Journal of Criminology*, 46: 935–49.

Newman, E. (2007) 'Weak States, State Failure, and Terrorism'. *Terrorism and Political Violence*, 19: 463–88.

Newman, O. (1972) *Defensible Space: Crime Prevention through Urban Design*. New York: Macmillan.

The News (2008) 'Guard Stuns Couple with Ban on Photos'. Retrived 18 February 2008, from http://www.portsmouth.co.uk/news/Guard-stuns-couple-with-ban.3632767.jp.

Nicholas, S., C. Kershaw and A. Walker (2007) *Crime in England and Wales 2006–7*. *Home Office Statistical Bulletin*. Retrieved 8 November 2007, from http://www.homeoffice.gov.uk/rds/pdfs07/hosb1107.pdf.

Noaks, L. (2000) 'Private Cops on the Block: a Review of the Role of Private Security in Residential Communities'. *Policing and Society*, 10: 143–61.

O'Conner, D., R. Lippert, K. Greenfield and P. Boyle (2004) 'After the "Quiet Revolution": the Self-Regulation of Ontario Contract Security Agencies'. *Policing and Society*, 14: 138–57.

O'Conner, D., R. Lippert, D. Spencer and L. Smylie (2008) 'Seeing Private Security Like a State'. *Criminology and Criminal Justice*, 8: 203–26.

O'Conner, M. (2007) *Fathers 4 Justice: the Inside Story*. London: Weidenfeld and Nicolson.

Oc, T. and S. Tiesdell (1997) 'Opportunity Reduction Approaches to Crime Prevention', in T. Oc and S. Tiesdell (eds) *Safer City Centres – Reviving the Public Realm*. London: Paul Chapman Publishing.

Osbourne, D. and T. Gaebler (1993) *Reinventing Government*. New York: Plume.

Oxley, J.C. (1993) 'Non-Traditional Explosive: Potential Detection Problems', in P. Wilkinson (ed.) *Technology and Terrorism*. London: Frank Cass.

Panorama (2008) *Britain's Protection Racket*. Retrieved 4 March 2008, from http://news.bbc.co.uk/1/hi/programmes/panorama/7195775.stm.

Parfomak, P.W. (2004) *Guarding America: Security Guards and the US Critical Infrastructure Protection*. CRS Report for Congress. Retrieved 20 August 2007, from http://www.fas.org/sgp/crs/RL32670.pdf.

Park, H. (2006) 'Defensible Parking Facilities for High-Rise Housing: a Study of South Korea'. PhD Thesis, University of Portsmouth.

Parsons, T. (1951) *The Social System*. London: Routledge and Kegan Paul.

Pearce, J. (2003) *Social Enterprise in Anytown*. London: Calouste Gulbenkian Foundation.

Perrow, C. (1984) *Normal Accidents: Living with High-Risk Technologies*. New York: Basic Books.

Phillips, S. and R. Cochrane (1988) 'Crime and Nuisance in the Shopping Centre: a Case Study in Crime Prevention'. Crime Prevention Unit Series Paper 16. London: Home Office.

Police Foundation and Policy Studies Institute Independent Inquiry (PF/PSI) (1996) *The Role and Responsibilities of the Police*. London: PF/PSI.

Porter, M.E. (2004) *Competitive Advantage*. New York: Free Press.

Poyser, S. (2003), 'From "Casbah" to "Quays" – Designing Crime in or Out?' MSc Dissertation, University of Portsmouth.

Prenzler, T. and R. Sarre (1999) 'A Survey of Security Legislation and Regulatory Strategies in Australia'. *Security Journal*, 12: 7–17.

Prenzler, T., T. Baxter and R. Draper (1998) 'Special Legislation for the Security Industry: a Case Study'. *International Journal of Risk, Security and Crime Prevention*, 3: 21–33.

Prison Officers Association (2006) *Public Services Not For Profit Campaign*. Retrieved 6 September 2007, from http://www.poauk.org.uk/campaign_NCmessage.htm.

Prison Officers Association (1997) 'Memorandum of Evidence', in Home Affairs Committee. *The Management of the Prison Service (Public and Private)*. London: The Stationery Office.

Private Security Authority (2007) *Private Security Services Employee Licence*. Retrieved 14 March 2008, from http://www.psa.gov.ie/psa/psa.nsf/Employee%20Booklet%20April%2007%20PSA%2011.pdf.

Private Security Authority (n.d.) *Licensing Information*. Retrieved 14 March 2008, from http://www.psa.gov.ie/psa/psa.nsf/agentvw?Openform&vw=psaLicensing.

Professional Security (2002) *Swedish Model*. Retrieved 20 August 2007, from http://www.professionalsecurity.co.uk/newsdetails.aspx?NewsArticleID=471&imgID=2.

Raab, J. and H.B. Milward (2003) 'Dark Networks as Problems'. *Journal of Administration Research and Theory*, 13: 413–39.

Raco, M. (2003) 'Remaking Place and Securitising Space: Urban Regeneration and the Strategies, Tactics and Practices of Policing in the UK'. *Urban Studies*, 40: 1869–87.

Ralph, R. (2004) 'The Carrying of Handcuffs by Security Officers in the United Kingdom: an Analysis of the Training and Legal Implications'. BSc Dissertation, University of Portsmouth.

Reeve, A. (1998) 'The Panopticisation of Shopping: CCTV and Leisure Consumption', in C. Norris et al. (eds) *Surveillance, Closed Circuit Television and Social Control*. Aldershot: Ashgate.

Reiner, R. (1978) *The Blue Coated Worker*. Cambridge: Cambridge University Press.

Reiner, R. (2000) *The Politics of the Police*. Oxford: Oxford University Press.

Reiner, R. (2005) 'Be Tough on Crucial Causes of Crime – Neoliberalism'. *Guardian Unlimited*. Retrieved 4 October 2007, from http://www.guardian.co.uk/ comment/story/0,,1649254,00.html.

Reiner, R. (2007) *Law and Order*. Cambridge: Polity.

Reiss, A.J. (1998) *Private Employment of Public Police*. Washington DC: National Institute of Justice.

Reppetto, T. (1976) 'Crime Prevention and the Displacement Phenomenon'. *Crime and Delinquency*, 22: 166–77.

Richardson, L. (2008) *DIY Community Action*. Bristol: Policy Press.

Rigakos, G.S. (2002) *The New Parapolice*. Toronto: University of Toronto Press.

Rigakos, G.S. and C. Yeung (2006) 'Canada', in T. Jones and T. Newburn (eds) *Plural Policing*. Abingdon: Routledge.

Riley, D. and M. Shaw (1985) *Parental Supervision and Juvenile Delinquency*. Home Office Research Study 83. London: HMSO.

Road Alert (1997) *Road Raging: Top Tips for Wrecking Roadbuilding*. Newbury: Road Alert.

Royal Society Study Group (1992) *Risk Analysis, Perception and Management*. London: The Royal Society.

Russek, H.I. and B.L. Zohman (1958) 'Relative Significance of Heredity, Diet, and Occupational Stress in Coronary Heart Disease of Young Adults'. *American Journal of Medicine*, 32: 266–75.

Sarre, R. and T. Prenzler (2005) *The Law of Private Security in Australia*. Pyrmont: Thomson.

Schneider, R.H. (2006) 'Contributions of Environmental Studies to Security', in M. Gill (ed) *The Handbook of Security*. Basingstoke: Palgrave.

Scott, T.M. and M. McPherson (1971) 'The Development of the Private Sector of the Criminal Justice System'. *University of Chicago Law Review*, 6: 267–88.

Secured Environments (2007) *Secured Environments is Based upon Six Principles*. Retrieved 5 May 2008, from http://www.securedenvironments.com/principles/index.aspx.

Security Industry Authority (SIA) (2004) *Annual Report and Accounts 2003–2004*. London: SIA.

Security Industry Authority (2006) *How to become an SIA Approved Contractor*. Retrieved 30 November 2006, from http://wwww.the-sia.org.uk/NR/rdonlyres/B2B71746-FA73-4039-8BBA-B012FEC78BF7/0/sia_acs_how_to.pdf.

Security Industry Authority (2007) *Stakeholder Engagement Strategy*. London: Security Industry Authority.

Security Industry Authority (n.d.a) *Get Licensed SIA Licensing Criteria*. Retrieved 12 November 2007, from http://www.the-sia.org.uk/NR/rdonlyres/725E43BE-2163-4E85-9151-6EAB15990BC1/0/sia_get_licensed.pdf.

Security Industry Authority (n.d.b) *Security Guarding – Required Training*. Retrieved 4 December 2007, from http://www.the-sia.org.uk/home/licensing/security_guarding/training/training_sg.htm.

Security Institute (2007) *Security Institute Yearbook 2007*. Nuneaton: Security Institute.

Security on Campus (2001) *What Jeanne Didn't Know*. Retrieved 22 October 2007, from http://www.securityoncampus.org/aboutsoc/didntknow.html.

Security Park (n.d.) *How Big is the Security Market?* Retrieved 5 December 2007, from http://www.securitypark.co.uk/security-market.asp.

Security Watchdog (2007) *Services: on Site Audit*. Retrieved 10 August 2007, from http://www.securitywatchdog.org.uk/art.php?art=8.

Shaftoe, H. and T. Read (2005) 'Planning Out Crime: the Appliance of Science or an Act of Faith?' in N. Tilley (ed.) *Handbook of Crime Prevention and Community Safety*. Cullompton: Willan.

Sharp, D. and D. Wilson (2000) '"Household Security": Private Policing and Vigilantism in Doncaster'. *Howard Journal*, 39: 113–31.

Shearing, C.D. (1992) 'The Relation between Public and Private Policing', in M. Tonry and N. Morris (eds) *Modern Policing*, volume 15. Chicago: University of Chicago Press.

Shearing, C.D. (1993) 'Policing: Relationships between Public and Private Forms', in M. Findlay and U. Zvekic (eds) *Alternative Policing*. Boston: Kluwer Law and Taxation Publishers.

Shearing, C.D. (2004) 'Thoughts on Sovereignty'. *Policing and Society*, 14: 5–12.

Shearing, C.D. (2006) 'Reflections on the Refusal to Acknowledge Private Government', in J. Wood and B. Dupont (eds) *Democracy, Society and the Governance of Security*. Cambridge: Cambridge University Press.

Shearing, C.D. and P.C. Stenning (1981) 'Modern Private Security: its Growth and Implications', in M. Tonry and N. Morris (eds) *Crime and Justice: an Annual Review of Research*. Volume 3. Chicago: University of Chicago Press.

Shearing, C.D. and P.C. Stenning (1982) 'Snowflakes or Good Pinches? Private Security's Contribution to Modern Policing', in R. Donelan (ed.) *Maintenance of Order in Society*. Ottawa: Ministry of Supply and Services.

Shearing, C.D. and P.C. Stenning (1987) 'Say "Cheese!": the Disney Order that is not so Mickey Mouse', in Shearing, C.D. and P.C. Stenning (eds) *Private Policing*. Newbury Park: Sage.

Shearing, C.D. and J. Wood (2003) 'Nodal Governance, Democracy and the New "Denizens"'. *Journal of Law and Society*, 30: 400–19.

Shearing, C.D., P.C. Stenning and S.M. Addario (1985) 'Police Perceptions of Private Security'. *Canadian Police College Journal*, 9: 127–53.

Shuman, E. (2001) 'El Al's Legendary Security Measures Set Industry Standards'. *Israel Insider*. Retrieved 26 October 2007, from http://www.israelinsider.com/channels/security/articles/sec_0108.htm.

Silke, A. (1998) 'The Lords of Discipline: the Methods and Motives of Paramilitary Vigilantism in Northern Ireland'. *Low Intensity Conflict and Law Enforcement*, 7: 121–56.

Silke, A. (1999) 'Ragged Justice: Loyalist Vigilantism in Northern Ireland'. *Terrorism and Political Violence*, 11: 1–31.

Silke, A. and M. Taylor (2000) 'War without End: Comparing IRA and Loyalist Vigilantism in Northern Ireland'. *Howard Journal*, 39: 249–66.

Silva, J.A., B. Gregory, M.D. Leong and R. Weinstock (1993) 'The Psychotic Patient as a Security Guard'. *Journal of Forensic Sciences*, 38: 1436–40.

Simonsen, C.E. (1996) 'The Case For: Security Management is a Profession'. *International Journal of Risk, Security and Crime Prevention*, 1: 229–32.

Singh, A. (2005) 'Private Security and Crime Control'. *Theoretical Criminology*, 9: 153–74.

Sklansky, D.A. (2006) 'Private Police and Democracy'. *American Law Review*, 43: 89–105.

Skogan, W. (1990) *Disorder and Decline: Crime and the Spiral of Decay in American Neighbourhoods*. New York: Free Press.

Smith, G. (2006) *The Fraud Problem*. Portsmouth: Institute of Criminal Justice Studies, University of Portsmouth.

Smith Institute (2007) *The Role of Business in Social Change: a Review of the Young Offender Programme Led by National Grid*. London: Smith Institute. Retrieved 7 December 2007, from http://www.smith-institute.org.uk/pdfs/the-role-of-business-review-of-young-offender-prog.pdf.

Spaninks, L., L. Quinn and J. Byrne (1999) *European Vocational Training Manual for Basic Guarding December 1999*. Retrieved 27 August 2007, from http://www.coess.org/documents/training_manual_en.pdf.

Spitzer, S. and A.T. Scull (1977) 'Privatisation and the Capitalist Development of the Police'. *Social Problems*, 25: 18–29.

Spurgeon, A., J.M. Harrington and C.L. Cooper (1997) 'Health and Safety Problems Associated with Long Working Hours: a Review of the Current Position'. *Occupational and Environmental Medicine*, 54: 367–75.

Stenning, P.C. (1989) 'Private Police and Public Police: Toward a Redefinition of the Police Role', in D.J. Loree (ed.) *Future Issues in Policing Symposium Proceedings*. Ottawa: Canadian Police College.

Stenning, P.C. (1994) 'Private Policing – Some Recent Myths, Developments and Trends', in D. Biles and J. Vernon (eds) *No 22: Private Sector and Community Involvement in the Criminal Justice System*. Proceedings of a conference held 30 November–2 December 1992, Wellington, New Zealand. Canberra: Australian Institute of Criminology.

Stenning, P.C. and C.D. Shearing (1979) 'Search and Seizure: Powers of Private Security Personnel'. Study Paper Prepared for the Law Reform Commission of Canada. Ottawa: Law Reform Commission of Canada.

Stenson, K. (2002) 'Community Safety in Middle England – the Local Politics of Crime Control', in G. Hughes and A. Edwards (eds) *Crime Control and the Community: the New Politics of Public Safety*. Cullompton: Willan.

Strange, S. (1996) *The Retreat of the State: the Diffusion of Power in the World Economy*. Cambridge: Cambridge University Press.

Strohm, C. (2004) *Covert Tests Reveal Airport Screening Failures*. Retrieved 20 August 2007, from http://www.govexec.com/dailyfed/0904/092204c1.htm.

The Sun Online (2004) *Airport Security Breached*. Retrieved 20 August 2007, from http://www.thesun.co.uk/article/0,,2-2004561139,00.html.

Sutton, A. (1994) 'Crime Prevention: Promise or Threat?' *Australian and New Zealand Journal of Criminology*, 27: 5–20.

Telegraph (2007) *Ratner Plans High Street Comeback*. Retrieved 14 May 2008, from http://www.telegraph.co.uk/money/main.jhtml?xml=/money/2007/10/14/cnr atner114.xml.

Thompson, T. and A. Wildavsky (1986) 'A Cultural Theory of Information Bias in Organisations'. *Journal of Management Studies*, 23: 273–86.

Tilley, N. (2005a) 'Crime Prevention and System Design', in N. Tilley (ed.) *Handbook of Crime Prevention and Community Safety*. Cullompton: Willan.

Tilley, N. (ed) (2005b) *Handbook of Crime Prevention and Community Safety*. Cullompton: Willan.

Timesonline (2007) *Norwich Union Fined Record £1.26 Million over Fraud Risk*. Retrieved 5 March 2008, from http://business.timesonline.co.uk/tol/business/industry_sectors/banking_and_finance/article3062076.ece.

Toft, B. and S. Reynolds (1997) *Learning from Disasters*. Leicester: Perpetuity Press.

Turner, B. (1978) *Man-made Disasters*. London: Wykeham.

United States Committee on Education and Labour (1971) *Private Police Systems*. New York: Arno Press and the New York Times.

United States Department of Transport (1993) *Audit of Airport Security Federal Aviation Administration*. Unpublished Report.

United States General Accounting Office (2001) *Aviation Security. Terrorist Acts Illustrate Severe Weaknesses in Aviation Security*. Washington DC: General Accounting Office.

Van Stedan, R. (2007) *Privatising Policing*. Amsterdam: BJU Legal Publishers.

Van Stedan, R. and R. Sarre (2005) 'The Growth of Private Security: Trends in Scandinavia and the European Union', paper presented to the European Society of Criminology conference, Krakow, 1 September.

Vila, B.J., D.J. Kenney, G.B. Morrison and M. Reuland (2000) *Evaluating the Effects of Fatigue on Police Patrol*. Retrieved 3 August 2007, from http://www.ncjrs.gov/pdffiles1/nij/grants/184188.pdf.

Waddington, P.A.J., D. Badger and R. Bull (2005) *The Violent Workplace*. Cullompton: Willan.

Wakefield, A. (2003) *Selling Security – the Private Policing of Public Space*. Cullompton: Willan.

Wakefield, A. (2006) 'The Security Officer', in M. Gill (ed.) (2006) *The Handbook of Security*. Basingstoke: Palgrave.

Walt, V. (2001) 'Unfriendly Skies are No Match for El Al'. *USA Today*. Retrieved 26 October 2007, from http://www.usatoday.com/news/sept11/2001/10/01/elal-usat.htm#more.

Wallis, R. (1993) 'Aviation Security', in P. Wilkinson (ed) *Technology and Terrorism*. London: Frank Cass.

Waterhouse, J.M., S. Folkard and D.S. Minors (1992) *Shiftwork, Heath and Safety: an Overview of the Scientific Literature 1978–1990*. London: HMSO.

Weatheritt, M. (1986) *Innovations in Policing*. Beckenham: Croom Helm.

Weber, M. (1948) 'Politics as a Vocation', in H.H. Gerth and C.W. Mills (eds and translators) *From Max Weber: Essays in Sociology*. London: Routledge and Keegan Paul.

Weber, T. (2002) *A Comparative Overview of Legislation Governing the Private Security Industry in the European Union*. Retrieved 22 August 2007, from http://www.coess.org/pdf/final-study.PDF.

Weir, D. (1996) 'Risk and Disaster: the Role of Communication Breakdown in Plane Crashes and Business Failure', in C. Hood and D. Jones (eds) *Accident and Design: Contemporary Debates in Risk Management*. London: UCL Press.

Weiss, R. (1978) 'The Emergence and Transformation of Private Detective and Industrial Policing in the United States, 1850–1940'. *Crime and Social Justice*, 1: 5–48.

Weitz and Luxenberg (2004) *The Lunatics Have Taken Over the Asylum*. Retrieved 6 November 2007, from http://www.weitzlux.com/premisesliability/news/hospitalmuder_1166.html.

Westall, A. (2001) *Value Led, Market Driven*. London: IPPR.

White, A. (forthcoming) 'The Re-Legitimation of Private Security in Britain'. PhD Thesis, University of Sheffield.

White House Security Review (1995) *Public Report of the White House Security Review*. Retrieved 5 May 2008, from http://www.fas.org/irp/agency/ustreas/usss/t1pubrpt.html.

Wilde, G. (1982) 'The Theory of Risk Homeostasis: Implications for Safety and Health'. *Risk Analysis*, 2: 209–25.

Wiles, P. and F. McClintock (eds) (1972) *The Security Industry in the United Kingdom*. Cambridge: Institute of Criminology, University of Cambridge.

Wilkinson, P. (2006) *Terrorism versus Democracy*. London: Routledge.

Williams, D., B. George and E. MacLennan (1984) *Guarding Against Low Pay – the Case for Regulation of Contract Security*. London: Low Pay Unit.

Wilson, J.Q. and G.L. Kelling (1982) 'Broken Windows: the Police and Neighbourhood Safety'. *Atlantic Monthly* (March): 29–38.

Wood, J. (2006) 'Dark Networks, Bright Networks and the Place of the Police', in J. Fleming and J. Wood (eds) *Fighting Crime Together: the Challenges of Policing and the Security Networks*. Sydney: University of New South Wales Press.

Wood, J. and C.D. Shearing (2007) *Imagining Security*. Cullompton: Willan.

Woodcock, J. (1994) *The Escape from Whitemoor Prison on Friday 9th September 1994*. Cm 2741. London: HMSO.

Yoshida, N. (1999) 'The Taming of the Japanese Private Security Industry'. *Policing and Society*, 9: 241–61.

Zedner, L. (2003a) 'Too Much Security?' *International Journal of the Sociology of the Law*, 31: 155–84.

Zedner, L. (2003b) 'The Concept of Security: an Agenda for Comparative Analysis'. *Legal Studies*, 23: 153–76.

Zedner, L. (2006) 'Liquid Security: Managing the Market for Crime Control'. *Criminology and Criminal Justice*, 6: 267–88.

Zimbardo, T. (1969) 'The Human Choice: Individuation, Reason and Order versus Deindividuation, Impulse and Chaos'. *Nebraska Symposium on Motivation*, 17: 237–307.

Zukin, S. (1995) *The Cultures of Cities*. Oxford: Blackwell.